National Climate Assessment Technical Report on the Impacts of Climate and Land Use and Land Cover Change

By Thomas Loveland, Rezaul Mahmood, Toral Patel-Weynand, Krista Karstensen, Kari Beckendorf, Norman Bliss, and Andrew Carleton

Open-File Report 2012–1155

U.S. Department of the Interior
U.S. Geological Survey

U.S. Department of the Interior
KEN SALAZAR, Secretary

U.S. Geological Survey
Marcia K. McNutt, Director

U.S. Geological Survey, Reston, Virginia: 2012

For more information on the USGS—the Federal source for science about the Earth,
its natural and living resources, natural hazards, and the environment—visit
http://www.usgs.gov or call 1–888–ASK–USGS

For an overview of USGS information products, including maps, imagery, and publications,
visit *http://www.usgs.gov/pubprod*

To order this and other USGS information products, visit *http://store.usgs.gov*

Suggested citation:
Loveland, Thomas, Mahmood, Rezaul, Patel-Weynand, Toral, Karstensen, Krista, Beckendorf, Kari, Bliss, Norman,
and Carleton, Andrew, 2012, National climate assessment technical report on the impacts of climate and land use and
land cover change: U.S. Geological Survey Open-File Report 2012–1155, 87 p.

Contents

Figures

Tables

Conversion Factors

SI to Inch/Pound

Multiply	By	To obtain
Length		
centimeter (cm)	0.3937	inch (in.)
millimeter (mm)	0.03937	inch (in.)
meter (m)	3.281	foot (ft)
kilometer (km)	0.6214	mile (mi)
meter (m)	1.094	yard (yd)
Area		
hectare (ha)	2.471	acre
square kilometer (km^2)	247.1	acre
hectare (ha)	0.003861	square mile (mi^2)
square kilometer (km^2)	0.3861	square mile (mi^2)
Volume		
cubic meter (m^3)	264.2	gallon (gal)
cubic centimeter (cm^3)	0.06102	cubic inch (in^3)
cubic kilometer (km^3)	0.2399	cubic mile (mi^3)
cubic meter (m^3)	0.0008107	acre-foot (acre-ft)
cubic hectometer (hm^3)	810.7	acre-foot (acre-ft)

Temperature in degrees Celsius (°C) may be converted to degrees Fahrenheit (°F) as follows:
°F=(1.8×°C)+32
Temperature in degrees Fahrenheit (°F) may be converted to degrees Celsius (°C) as follows:
°C=(°F-32)/1.8
Horizontal coordinate information is referenced to the insert datum name (and abbreviation) here, for instance, "North American Datum of 1983 (NAD 83)"

National Climate Assessment Technical Report on the Impacts of Climate and Land Use and Land Cover Change

By Thomas Loveland[1], Rezaul Mahmood[2], Toral Patel-Weynand[3], Krista Karstensen[1], Kari Beckendorf[4], Norman Bliss[4], and Andrew Carleton[5]

Abstract

This technical report responds to the recognition by the U.S. Global Change Research Program (USGCRP) and the National Climate Assessment (NCA) of the importance of understanding how land use and land cover (LULC) affects weather and climate variability and change and how that variability and change affects LULC. Current published, peer-reviewed, scientific literature and supporting data from both existing and original sources forms the basis for this report's assessment of the current state of knowledge regarding land change and climate interactions. The synthesis presented herein documents how current and future land change may alter environment processes and in turn, how those conditions may affect both land cover and land use by specifically investigating,

- The primary contemporary trends in land use and land cover,
- The land-use and land-cover sectors and regions which are most affected by weather and climate variability,
- How land-use practices are adapting to climate change,
- How land-use and land-cover patterns and conditions are affecting weather and climate, and
- The key elements of an ongoing Land Resources assessment.

These findings present information that can be used to better assess land change and climate interactions in order to better assess land management and adaptation strategies for future environmental change and to assist in the development of a framework for an ongoing national assessment.

1.0 Introduction

The 2003 Climate Change Science Program (CCSP) science strategy (Climate Change Science Program, 2003) initially recognized the complex links between LULC and weather and climate as one of the primary global climate change science themes. The CCSP LULC science strategy focused on the following question: How do climate variability and change affect land use and land cover, and what are the potential feedbacks of changes in land use and land cover to climate?

The objective of this assessment is to understand how current and predicted changes in land cover and land use will alter weather, climate, and other global environmental conditions and in turn, how those conditions will change land cover and land use. This interaction is a necessary basis for managing and adapting to future environmental changes.

---

[1] U.S. Geological Survey, Earth Resources and Observation Science Center, Sioux Falls, South Dakota.
[2] Western Kentucky University, Bowling Green, Kentucky.
[3] U.S. Forest Service, Research and Development, Quantitative Sciences, Arlington, Virginia.
[4] ASRC Research and Technology Solutions, on contract to U.S. Geological Survey, Earth Resources and Observation Science Center, Sioux Falls, South Dakota.
[5] Pennsylvania State University, University Park, Pennsylvania.

This report and the land use and land cover workshop report (Lebow and others, 2011) are designed to supplement the state-of-knowledge of the current connections between LULC and weather and climate and to evaluate the general readiness of the Federal USGCRP agencies and broader science community to carry out an ongoing assessment of LULC and weather and climate impacts. The respective convening Federal sponsors, the U.S Forest Service (workshop report) and U.S. Geological Survey (technical report), consider both documents as complementary volumes for an understanding of the current LULC foundation for addressing weather and climate issues.

Centralizing around LULC in a land-change-science context, this report focuses on the state-of-knowledge of climate impacts on LULC for the 2009–12 period in an effort to assess LULC and weather and climate interactions within the NCA regions, including those affecting natural resources. Land-change science, as defined by Turner and others, 2007, stresses the understanding of LULC dynamics within a coupled human-natural environment context that considers the causes, impacts, and consequences of LULC change.

1.1 Land Use and Land Cover Definitions and Concepts

For the purposes of this report, land use includes the human activities and management practices for which land is used; land cover includes the status of vegetation, bare soil, developed structures (for example, building, roads, and other infrastructure), and water bodies including wetlands. See appendix A for additional definitions of major terms used in this report. Land use and land cover are linked but the relations are complex. Land-use practices determine cover, and subsequently, the effects of cover on weather and climate. Land-cover changes affect the exchange of water, energy, and gases with the atmosphere because of changes in surface albedo and transpiration, which modify surface temperature and latent heat flux and ultimately can cause changes in regional temperature and precipitation patterns (Dale, 1997). Land-cover changes also can initiate a wide range of other global changes including water-resources alterations and modifications to habitat. It is equally important to recognize that the resulting effects of climate change can force changes in land use, and potentially land cover, as land-management practices are implemented that mitigate the adverse effects of weather and climate variability and change.

When considering LULC and weather and climate, several factors should be considered:
- LULC changes are local phenomena. However, the impacts of these changes can reach beyond the local scale and can accumulate and have global impacts. For example, tropical forest clearing reduces the ability to sequester carbon, which ultimately affects atmospheric chemistry worldwide.
- LULC changes are, to a certain extent, driven by natural events but are mostly attributable to human activities on the Earth.
- LULC change is typically motivated by economic goals, which are important for the growth of personal income and expanding the economies of communities, States, and nations. Because healthy economies depend on profitable land uses, mitigation strategies that reduce unwanted negative consequences must consider local to national economic implications.
- LULC change can influence land-management practices by altering ecosystem goods and services and is also the catalyst for changes in atmospheric chemistry, weather and climate variability, and other processes.

1.2 An Overview of Climate and Land Use/Cover Forcings and Feedbacks

Land use and land cover (LULC) play critical roles in land surface-atmosphere interactions and thus influence climate. These impacts are evident on local, regional, and global climate. In addition,

LULC has biogeophysical and biogeochemical impacts on climate. Biogeophysical mechanisms alter radiation balance, energy partitioning, and exchanges of energy, mass, and momentum between the land surface and the atmosphere.

The radiation balance equation may be written as follows (Oke, 1987):

$$Q^* = K^* + L^* \qquad (1)$$

where,

Q^*	is net all wave radiation,
K^*	is net short-wave radiation, and
L^*	is net long-wave radiation.

Over dry land areas (light-colored surfaces) more short-wave radiation is reflected and contrasted with land surfaces having relatively high soil-moisture content (dark-colored surfaces), where more short-wave radiation is absorbed. The surface-water content also influences the net radiation balance.

The energy balance relations can be expressed as follows (Oke, 1987):

$$Q^* = Q_E + Q_H + Q_G \qquad (2)$$

where Q_E, Q_H, and Q_G represent latent, sensible, and ground energy fluxes, respectively.

On land surfaces with higher soil-moisture content most of the available energy is partitioned into latent as opposed to sensible flux. Higher latent energy flux (higher evapotranspiration) typically leads to lowering of temperature and increases atmospheric humidity, convection, and cloud development. Over drier land areas, such as semiarid or arid lands, sensible energy flux (surface and atmospheric warming) dominates the energy partitioning. Seasonal patterns, such as wet and dry seasons, influence the ways that LULC interacts with climate.

These impacts and their interactions reach a higher level of complexity because of the heterogeneity of the land surface, overlying vegetation cover, and vegetation condition. The types of vegetation cover (grass compared to tree, for example) alter the energy partitioning (lower compared to higher latent energy flux). For example, given similar soil water and energy availability, there is more evapotranspiration from tree-covered land surfaces (greater latent energy flux) compared to grass-covered land surfaces. These partitioning differences also are dependent, particularly on plant physiology and their response to environmental stress. These may include excessively dry or unusually wet conditions. Vegetation type, stand architecture, and transition from one type of vegetation cover to another (for example, grass to forest) affect the transfer of energy and mass to the atmosphere and hence, convection and cloud development. Moreover, climate is not a passive player in the feedback with LULC. Intraannual and interannual climate variability (for example, dry compared to wet; healthy compared to stressed vegetation cover) affects the intensity of LULC forcing and the above mentioned processes and phenomena which, in turn, affect the atmospheric state and related phenomena. In some regions, this feedback loop is extended even at interannual time-scales.

Climate changes also affect LULC through direct effects on vegetation growth and structure and indirectly through human adaptation to climate change. For example, droughts may cause water stress in vegetation that may lead to more surface heating and less evapotranspiration. The stressed vegetation may be subject to fire, which will influence the radiation balance immediately after the fire, and perhaps vegetation with a different albedo will regrow. If drought causes crop failures for many years, an agricultural land use may be abandoned and a transition may follow to a grass land cover. The mutual

3

interactions of land use and climate may be described as a feedback loop with each system influencing the other. The feedbacks may be very complex and occur at various scales in space and time.

LULC changes can modify the interactions between land surfaces and the atmosphere. For example, with replacement of forests by barren surfaces, agriculture, or urban areas, an existing land surface-atmosphere feedback loop may be substituted with a modified loop. Land-use changes within a particular land-cover type may further complicate the ability to understand and describe the forcings and feedbacks. Replacement of rain fed agriculture (which already replaced natural grasslands) with irrigated agriculture is an example of such a change. In some cases, single-crop land use (one crop per year) has changed to multicrop land use (three crops per year). These types of changes may alter local and regional scale climate. Impacts at the global scale remain a topic of further investigation.

2.0 State of Knowledge

2.1 Current, Past, and Future Land-Cover Understanding

LULC connections with weather and climate require general knowledge of past LULC patterns and changes, detailed data and information on current LULC patterns and dynamics, and an understanding of the plausible scenarios and resulting LULC patterns associated with change 20–100 years into the future. To the extent possible, spatially explicit LULC products are used because they offer the ability to evaluate the geographic variability and complexity of climate and LULC impacts. A more complete summary of LULC products can be found in "Land Use and Land Cover National Stakeholder Workshop Technical Report" (Lebow and others, 2011).

2.1.1 Current Land Cover in the United States

The two most common spatially explicit maps of current land cover in the United States are the 2001 and 2006 National Land Cover Database (NLCD) products (Homer and others, 2004; Fry and others, 2011), and the U.S. Forest Service (USFS)-U.S. Geological Survey (USGS) 2001 LANDFIRE land cover and vegetation dataset (Vogelmann and others, 2011). LANDFIRE provides detailed vegetation type and structure data sets for the entire United States, including Alaska and Hawaii, for use in fire management activities. The products are based on the National Vegetation Classification System and were produced through the analysis of Landsat imagery from 2001 and 2008. The 2008 vegetation updates are nearing completion. LANDFIRE datasets emphasize natural vegetation rather than general land cover. As a result, for assessments of LULC, LANDFIRE vegetation layers are best suited to assessments of climate impacts and feedbacks involving natural landscapes.

The NLCD 2001 and 2006 land cover databases use Landsat 5 and 7 imagery to produce 30 meter (m) resolution maps of land cover types, percent tree cover, and percent urban imperviousness. NLCD 2001 covers all 50 states, while the 2006 NLCD is limited to the conterminous United States. NLCD 2006 also quantifies land cover change between the years 2001 to 2006. A formal accuracy assessment is part of each dataset, though the NLCD 2006 accuracy assessment will not be completed until 2012.

Because of the differences in content and coverage, the NLCD 2001 land-cover map provides the most comprehensive summary of land-cover composition in the United States (fig. 1). Based on 2001 national land-cover statistics, shrub/scrub vegetation (including dwarf shrubs, moss, and lichen cover in Alaska) covers 39.2 percent of the United States, trees cover 23.2 percent, agriculture (cropland and pasture) covers 18.6 percent, open water covers 6.5 percent, developed and built-up lands cover 4.0

percent, barren lands represent 2.6 percent, and snow and ice cover 0.9 percent of the 50 States. Table 1 provides detailed land-cover statistics for the Nation and each of the eight NCA regions.

Table 1. 2001 land-cover percentages for the United States and eight National Climate Assessment regions (Fry and others, 2011).

Land cover class	Northeast	Southeast	Midwest	Great Plains	Southwest	Northwest	Alaska	Hawaii	U.S.
Low intensity residential	4.1	4.8	4.6	2.7	1.4	1.7	0	3.1	2.4
High intensity residential	3.1	2.1	2.4	0.9	0.7	0.9	0.1	2.3	1.1
Commercial/ Industrial/ Transportation	1.8	0.6	0.7	0.3	0.5	0.3	0	0.8	0.4
Developed High Intensity	0.6	0.2	0.3	0.1	0.1	0.1	0	0.5	0.1
Pasture/Hay	5.5	11.2	11.6	7	1.3	2.3	0	2.2	5.2
Row crops	5.4	11.8	37.4	22.7	4.3	8.2	0	2.6	13.4
Dwarf shrub	0	0	0	0	0	0	17.2	0	3.4
Shrub/Scrub	3	4.2	0.5	16	50.4	34.8	21.2	22.7	22
Grassland/ Herbaceous	0.4	3.6	2.4	33.6	15.4	7.9	0.8	10.5	12.7
Sedge/Herbaceous	0	0	0	0	0	0	5.7	0	1.1
Lichens	0	0	0	0	0	0	0	0	0
Moss	0	0	0	0	0	0	0	0	0
Deciduous forest	22.7	18.8	20.2	3.6	2	0.7	3.4	0	7.6
Evergreen forest	13.3	15.5	2.5	6.5	17.1	35.4	15.5	22	13.4
Mixed forest	16.4	4.5	0.9	0.5	0.8	1.6	3.5	0	2.2
Woody wetlands	6.2	12	4.2	1.7	0.4	0.6	3.5	0.2	3.4
Emergent herbaceous wetlands	1.8	3.2	1.6	1	0.3	0.7	2.9	0.1	1.6
Open water	14.9	7.3	10.4	1.9	1.7	3.1	14.2	21.7	6.5
Perennial Ice/Snow	0	0	0	0	0	0.1	4.3	0	0.9
Bare Rock/ Sand/Clay	0.8	0.3	0.2	0.5	3.7	1.5	7.7	11.2	2.6
	100	100	100	100	100	100	100	100	100

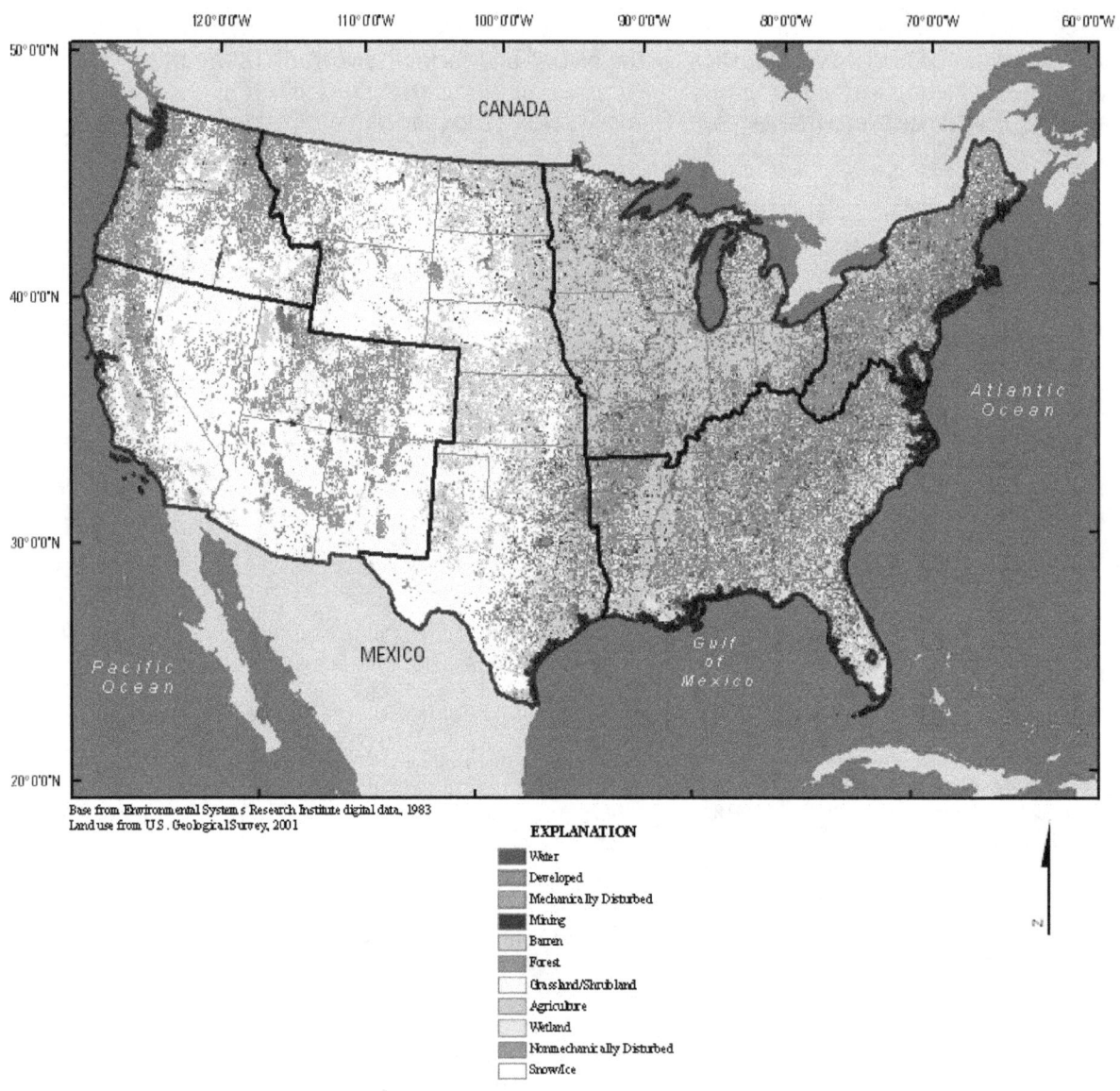

Figure 1. Land-cover patterns and National Climate Assessment regions for 2001 (Homer and others, 2004).

2.1.2 Past Land Cover

Accurate portrayal of past LULC is especially challenging. While the 2003 CCSP LULC change science strategy recommended development of a complete LULC history in the United States for use in climate research, progress has been limited. Several Federal agencies within the U.S. Department of Agriculture (USDA), including the Natural Resources Conservation Service and the Economic Research Service, have relatively systematic records of past land uses. Others have census or inventory information on particular sectors (Nusser and Goebel, 1997; Gillespie, 1999; National Agricultural

Statistical Service, 2009). The USGS has several research LULC products that provide more information on past LULC (Waisanen and Bliss, 2002 for a summary of 1790 to 1997 changes in U.S. population and agriculture; Loveland and others, 2002 for a strategy for monitoring contemporary U.S. land cover change; Fry and others, 2011 information on 2001–06 U.S. land cover change). All of these USGS historical datasets cover the conterminous United States but coverage often does not include Alaska and Hawaii.

An important characteristic of historical LULC change in the United States is the considerable geographic variability in the types, rates, and causes of change. The USGS Land Cover Trends study provides an assessment of the 1973 to 2000 variability. Historical Landsat imagery and probability-based sampling were used to estimate change for each of the 84 conterminous ecoregions in the United States (Loveland and others, 2002). Based on this study, an estimated 8.6 percent of the conterminous United States changed from 1973 to 2000, but the ecoregional change varied from less than 1 percent in the southwest desert ecoregions to over 30 percent for several northwest and south-central forested ecoregions. Using the trends data, change rates for six conterminous NCA regions in the United States were calculated (table 2). For the 1973 to 2000 period, the Southeast NCA region had the highest rate of change (16.7 percent) because of active forest timber harvesting and replanting, and the Southwest NCA had the lowest rate of change (4.0 percent). The following text is a summary of land change in the six NCA regions. The order of the summary is from highest to lowest rates of NCA regional change. See appendix B for definitions of the land-cover classes discussed in the following sections.

Table 2. Overall 1973–2000 land-cover change for the National Climate Assessment (NCA) regions (U.S. Geological Survey, 2012).

Region	Percent Change
National	8.6
Southeast	16
Northwest	12.7
Great Plains	8.7
Northeast	6.9
Midwest	5.7
Southwest	4

2.1.2.1 Southeast

The Southeast region recorded the highest amount of land-cover change with 16 percent of the region changed from one land cover type to another (for example, agriculture to forest cover). To understand land-cover dynamics, it is important to understand both gross and net change. Gross change is the total amount of a given land cover changed, and net change is the aggregate of gains and losses affecting that category. For example, the Southeast experienced a 7.9 percent gross loss of forest cover but the net loss was only 2.5 percent. This is largely the result of active management (planting and harvesting rotations) of pine plantations (Drummond and Loveland, 2010).

Overall, Southeast forest land cover declined from 48.3 percent of the region in 1973 to 45.8 percent in 2000. Agriculture was the second leading land cover in the region (25.6 percent) and also declined from 1973 levels. The third leading land cover in the region was wetland with a decline from 11.4 percent in 1973 to 10.7 percent in 2000. The major increases in cover were in the developed (urban and built-up) and mechanically disturbed land-cover classes, which increased steadily from 1973 to 2000.

Regarding gross land-cover change, forest cover (7.9 percent), mechanical disturbances (3.9 percent), agriculture (3.7 percent), developed (2.3 percent), and wetland (1.3 percent) experienced the largest gross changes. The leading land cover transitions occurring in the Southeast region were:
- Forest cover to mechanically disturbed
- Agriculture to forest
- Forest cover to developed
- Mechanically disturbed to forest, and
- Forest to agricultutre

Appendix C provides tabular summaries of Southeast land-cover change statistics.

2.1.2.2 Northwest

Overall land-cover change was 12.7 percent between 1973 and 2000 (table 2). Forest cover declined from 42.7 percent of the region in 1973 to 40.3 in 2000 because of wildfires and other disturbances, urbanization, and active forest management (for example, harvesting and replanting). Nonmechanical disturbances (primarily wildfire disturbances in this region) increased by nearly 1.8 percent, and developed cover increased by 0.5 percent (from 1.2 to 1.7 percent). The major land-cover conversions in the region were:
- Forest to grassland/shrubland
- Forest to mechanically disturbed
- Mechanically disturbed to forest
- Forest to nonmechanically disturbed, and
- Agriculture to grassland/shrubland

The Northwest land-cover classes that experienced the largest gross change were: forest (6.0 percent), grassland/shrubland (4.6 percent), mechanically disturbed (2.7 percent), nonmechanically disturbed (2.1 percent), and agriculture (1.9 percent). Appendix D provides detailed change statistics for this region.

2.1.2.3 Great Plains

Overall land change in the Great Plains between 1973 and 2000 was 8.7 percent (table 2). The major changes were a 1.6-percent loss of agricultural cover and a nearly equal gain in grassland/shrubland cover. Developed cover increased by about 0.4 percent. The primary land cover conversions in the region were:
- Agriculture to grassland/shrubland
- Grassland/shrubland to agriculture
- Forest to grassland/shrubland
- Forest to mechanically disturbed, and
- Agriculture to developed

The major gross changes were: grassland/shrubland (5.3 percent), agriculture (4.8 percent), forest (1.5 percent), mechanically disturbed (0.50 percent), and water (0.46 percent). This region experienced several unique episodes of change beginning with a significant increase in agriculture from 1973 to 1986 and then an even larger decline from 1986 to 2000. The later period of change was driven by the Conservation Reserve Program objectives (Drummond and others, 2012).

The Great Plains NCA region illustrates the challenges in consistently summarizing and describing land-cover change. Some subregions, such as the High Plains (the subregion overlaying the High Plains aquifer) experienced much higher rates of change because of the intensification of irrigation, while other areas, such as the Flint Hills and Sand Hills of Nebraska, experienced very low rates of change because the land uses in those subregions had limited potential for more intensive land uses (Drummond and others, 2012). Appendix E provides detailed change statistics for this region.

2.1.2.4 Northeast

The overall land cover change in the Northeast between 1973 and 2000 was 6.9 percent (table 2). The two major changes were a 2.2 percent decrease in forest cover and a 1.4 percent increase in developed land. Grassland/shrubland cover increased 0.7 percent while agricultural cover decreased 0.9 percent. The leading land cover conversions in the region were:

- Forest to mechanically disturbed
- Forest to developed
- Forest to grassland/shrubland
- Agriculture to developed, and
- Agriculture to forest

The major gross land-cover changes were: forest (4.1 percent), agriculture (1.5 percent), mechanically disturbed (1.4 percent), developed (1.4 percent), and grassland/shrubland (1.1 percent). for An analysis of land cover change in the Southeast and Northeast regions can be found in Drummond and Loveland (2010). Appendix F includes detailed statistics on Northeast land-cover changes.

2.1.2.5 Midwest

The overall amount of land cover change in the Midwest between 1973 and 2000 was 5.7 percent (table 2). The predominant land-cover changes were the expansion of developed lands by an additional 1.3 percent of the region and a 1.4-percent decline of agriculture within the region. Forest cover declined by approximately 0.9 percent. The major gross changes were: agriculture (2.7 percent), forest (2.3 percent), grassland/shrubland (1.4 percent), developed (1.4 percent), and mechanically disturbed (0.8 percent). The leading land cover transformations in the region were:

- Agriculture to developed
- Agriculture to grassland/shrubland
- Forest to mechanically disturbed
- Forest to agriculture, and
- Forest to grassland/shrubland

Appendix G includes includes detailed statistics on Midwest land-cover changes.

2.1.2.6 Southwest

The Southwest NCA region underwent the least amount of land-cover change with an overall amount of land change of 4.0 percent between 1973 and 2000 (table 2). Forest cover and agriculture decreased 0.5 and 0.4 percent, respectfully, and nonmechanical disturbances (for example, wildfire disturbances) and developed land each increased by 0.5 percent. The top leading land-cover conversions in the region were:

- Agriculture to grassland/shrubland
- Grassland/shrubland to agriculture
- Forest to nonmechanically disturbed
- Grassland/shrubland to developed, and
- Grassland/shrubland to nonmechanically disturbed

The main gross land cover changes were: grassland/shrubland (2.5 percent), agriculture (1.7 percent), forest (0.8 percent), nonmechanically disturbed (0.6 percent), and developed (0.5 percent). Appendix H has detailed land-cover change statistics for the Southwest.

2.1.2.7 Land Cover Change Since 2000

The USGS NLCD provides the most current spatially explicit information on land-cover change for the conterminous United States (Fry and others, 2011). The NLCD provides baseline land cover for 2006 and 2001–06 land-cover change data (figs. 2 and 3). At this time, the accuracy of the baseline and land-cover change products is unknown, but results will be released in 2012. Preliminary results show that the overall amount of change between 2001 and 2006 was 1.67 percent, with the greatest changes in forest cover with a decrease of approximately 1 percent and developed land cover with an increase of 0.4 percent (Fry and others, 2011).

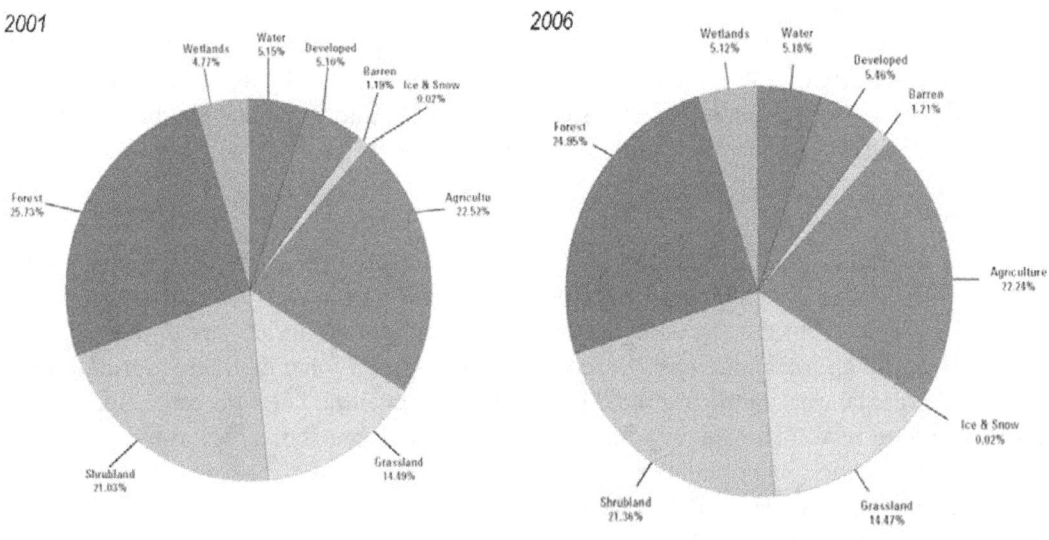

Figure 2.   National Land Cover Database (NLCD) statistics for 2001 and 2006 (Fry and others, 2011).

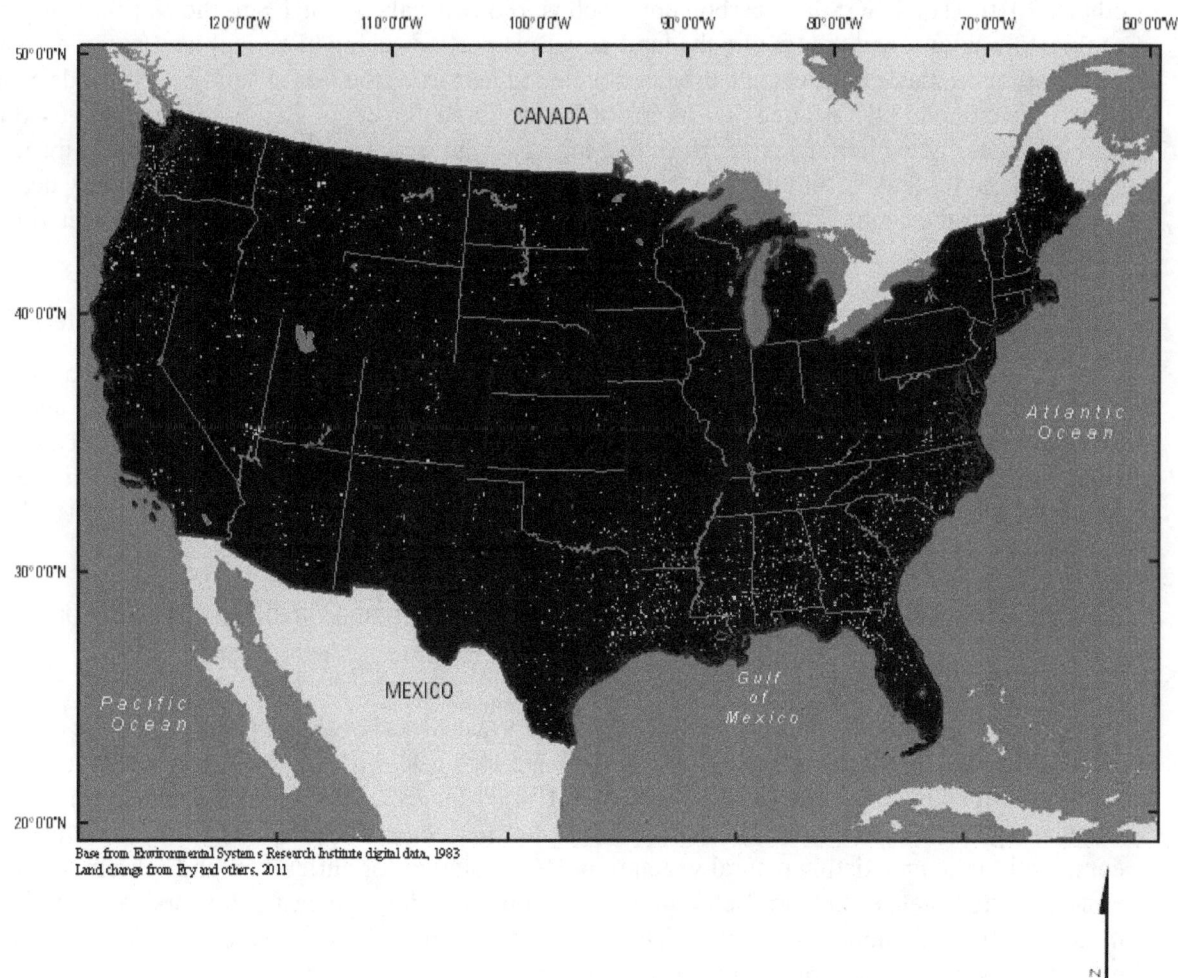

Figure 3. National Land Cover Database (NLCD) 2001–06 land-cover change. The light colored areas in this map represent areas of change that occurred between the 2 years (Fry and others, 2011).

2.1.3 Future Land Cover in the United States

Several agency products are designed to aid in the achievement of specific mandates or objectives. Two reports relevant for a LULC-climate assessment are from the U.S. Environmental Protection Agency (EPA) and USGS. Both are based on the Intergovernmental Panel on Climate Change (IPCC) Special Report on Emissions Scenarios (SRES) social, economic, and demographic storylines (Nakicenovic and Swart, 2000). The EPA has developed Integrated Climate and Land-Use Scenarios (ICLUS) based on four SRES and has used demographic model population estimates as a means to spatially distribute housing density across the conterminous United States for each scenario by decade to the year 2100 (U.S. Environmental Protection Agency, 2009). The EPA ICLUS emphasize built-up land uses rather than the broader set of resource-based land covers.

For a broader characterization, the USGS land carbon project is using a set of scenarios based on the SRES to produce land-cover forecasts for all major land-cover types in the United States (Zhu and

11

others, 2010). The USGS land carbon approach is to downscale the SRES to the Nation's major ecological regions, and then refine the land-cover proportions using historical land-cover data and information on major land-change drivers developed as part of the USGS land cover trends project. The USGS-developed FOREcasting SCEnarios of Future Land Cover (FORE–SCE) model is used to develop spatially explicit maps of projected land cover, by ecoregion, on an annual basis through 2100 (Sohl and Sayler, 2008). At this time, USGS scenarios and land-cover forecasts have only been completed for the central United States. The remaining regions are expected to be completed in late-2013.

2.2 Land-Use and Land-Cover Patterns and Conditions Affecting Weather and Climate

A number of studies were conducted over the last several decades highlighting the role of LULC change in global climate (Claussen and others, 2001; DeFries and others, 2002; Feddema and others, 2005; Brovkin and others, 2006; Davin and others, 2007; Pitman and others, 2009; Davin and Noblet-Ducoudré, 2010; Pongratz and others, 2010). However, a detailed discussion on global impacts is beyond the scope of this report. Here, the primary focus is on LULC change impacts on weather and climate in the United States .

2.2.1 Land Use and Land Cover Change Because of Agriculture and Changes in Weather and Climate

2.2.1.1 Planetary Boundary Layer

The Planetary Boundary Layer (PBL) is the lowest level of the troposphere and is directly affected by characteristics of the Earth's surface and changes in LULC. In continental interior regions of the Northern Hemisphere, especially the central United States, LULC change-driven, rainfed, and irrigated agriculture adoption have significantly modified regional-scale energy and moisture budgets compared to the preexisting natural vegetation. Deforestation for agriculture and settlement in the eastern United States and Corn Belt regions over approximately the past 150 years has significantly modified physical climate parameters. The reduction of canopy height has reduced the aerodynamic roughness, which affects momentum exchange and the convective fluxes of heat and moisture, between the land surface and atmosphere and eventually cloud development (Allard and Carleton, 2010; Matyas and Carleton, 2010; and Carleton and others, 2008a, b). A similar reduction in leaf-area index over the same period increased the surface albedo, or short-wave reflectance, particularly in the Midwest, and also increased the heterogeneity (patchiness) of the land cover. This increased patchiness has altered the regional and smaller-scale surface energy budgets. A previously based modeling study (McPherson and Stensrud, 2005) demonstrated that the wheat belt of western Oklahoma has modified meso-scale horizontal and vertical wind fields, potential temperature, height of the PBL (lowered), and overall meso-scale circulation.

Recent studies from the other parts of the world also report LULC change impacts on PBL. For example, a series of studies set in the Canadian Prairies clearly found increases in convective activity as a result of replacement of natural grassland by farming (Hanesiak and others, 2004; Raddatz, 2003; Raddatz and Cummine, 2003; Raddatz, 2007). Like many other regions of the world, Australia also experienced LULC change. It is found from observational studies that conversion of native vegetation cover to agriculture suppressed cloud development over modern vegetation cover; that is, native vegetation provided a favorable condition for cloud formation (Lyons and others, 1993; Lyons, 2002). Changes in surface roughness and albedo played an important role in this suppression. Moreover, this conversion of land cover also favors dust devil formation (Lyons and others, 2008). Sen Roy and others

(2011) reported changes in wind field and equivalent potential temperature because of irrigated agriculture.

2.2.1.2 Temperature

As noted earlier, LULC change also can modify air temperature and near-surface moisture content. A study based on observed data, (Fall and others 2010) found that conversion of lands (for all types) to agriculture resulted in cooling. It was noted that conversion of barren areas and grasslands/shrublands was associated with a cooling of -0.121 degrees Celsius (°C) and -0.096°C, respectively. In addition, conversion of previously forested and barren areas to agriculture resulted in -0.061°C and -0.039°C, respectively. Previously, Hale and others (2006, 2008) analyzed LULC change and temperature data and indicated LULC change may have played an important role in temperature changes over the United States. In an earlier study, Kalnay and Cai (2003) reported decreases of maximum temperature over the Midwestern United States during the spring, summer, and autumn seasons. These changes were linked to rainfed, agricultural, land-use change. McPherson and others (2004) found low anomalies of maximum temperatures in the observed data for the rainfed, winter wheat growing area of western Oklahoma and also reported higher dewpoint temperatures (T_d) over wheat growing areas compared to surrounding native grasslands in Oklahoma. These findings were further supported by Sandstrom and others (2004). Analysis of observed data suggested an increase in frequency of occurrence of days experiencing extreme T_d (greater than or equal to 22°C) in the central United States (Sandstrom and others, 2004) and have suggested that increased evapotranspiration from croplands (LULC change) is the primary cause of this dew-point increase.

Two critical studies by Bonan (1997, 2001) reported impacts of agriculture on United States temperatures. From a modeling study, Bonan (1997) reported up to 2°C cooling of summer temperature over the central United States and up to 1.5 grams/kilogram (g kg^{-1})increase in atmospheric moisture content in much of the United States. Bonan noted that these changes were caused by lowered surface roughness and stomatal resistance and increased albedo because of replacement of forests with modern vegetation (largely agriculture). In a subsequent study, Bonan (2001) analyzed near-surface temperatures in the United States and found lowering of the daily maximum temperature. A temporal correlation between the expansion of agriculture and lowered of temperature also was found by Bonan.

Impacts of agriculture-related changes of weather and climate also have been experienced in other regions of the world. Based on an observational study, Gameda and others (2007) found significant reductions in mid-June to mid-July maximum air temperatures, and diurnal temperature range and solar radiation of 1.7°C decade^{-1}, 1.1°C decade^{-1} and 1.2 M J m^{-2} decade, respectively, in the Canadian Prairies. These changes were linked to the increased latent heating associated with increased area under crop cultivation that forced lowering of temperature. In addition, observational data suggest -0.30°C cooling in semi-arid Almeira, Spain (Campra and others, 2008), associated with pastureland changing to greenhouse farming and resultant strong negative radiative forcing (up to -34 watts per square meter (W m^{-2})). To evaluate the impacts of historical land-cover changes on Australia's regional climate, McAlpine and others (2007) conducted two sets of model experiments comparing the climatic impacts of modern day (1990) compared to pre-European land-cover conditions. A modeling study by McAlpine and others (2009) found 0.2–2.0°C increase in summer temperature for eastern Australia and 0.5°C for southwestern western Australia as a result of LULC change. It was suggested that increased summer temperatures in eastern Australia were linked to a large decrease in vegetation fraction (19 percent) and leaf area index (LAI) (23 percent) (McAlpine and others, 2009).

Based on a modeling study, Beltrán-Przekurat and others (2011) found a large change (cooling) in maximum temperatures linked to LULC change in the Pampas region of Argentina. Simulations suggested large decreases in sensible energy flux and increases in latent energy flux. The study also noted a decrease in diurnal temperature range. The results were comparable to the observed trends in temperature over Argentina (Rusticucci and Barrucand, 2004; Nuñez and others, 2008).

Modification of LULC to irrigated agriculture further intensified some of these impacts. For example, analyses of observed data suggest up to 1.41°C lowering (cooling) of maximum temperatures during the post-1945 period for the growing season (May through September) over irrigated locations in Nebraska (Mahmood and others, 2006). This finding is supported by the modeling studies by Mahmood and Hubbard (2002, 2004), Adegoke and others (2003), and Ozdogan and others (2010); the authors show up to a 36-percent increase in evapotranspiration over irrigated areas compared to grass. As discussed previously, increased evapotranspiration (latent energy flux) lowers sensible energy flux and thus temperature. In an effort to further verify these results, Mahmood and others (2008) investigated the impacts of irrigation on observed near-surface atmospheric moisture content using long-term observed T_d data for Nebraska and found a 1.56°C increase in average growing season T_d over irrigated areas. A 2.17°C increase in T_d for peak growing season months also was reported (Mahmood and others, 2008). Earlier, Mahmood and others (2004) reported a cooling trend in long-term extreme maximum temperatures for irrigated locations in the same region.

Similar studies on impacts of irrigation also reported a lowering of growing season temperature in California (Christy and others, 2006; Lobell and others, 2006a,b; Bonfils and Lobell, 2007; Kueppers and others, 2007, 2008; Lobell and Bonfils, 2008; Jin and Miller, 2011; Sorooshian and others, 2011). For example, based on a modeling study, Sorooshian and others (2011) reported 3–7°C cooling and 9–20 percent increase in relative humidity over irrigated areas. This study also suggests that irrigated land use could lower daily maximum temperatures up to 1–4°C over surrounding nonirrigated areas. Jin and Miller (2011) reported cooling of daily maximum temperature during all seasons for irrigated locations with 0.30°C/decade reduction over summer. This modeling study further explained the overall findings.

Gordon and others (2005) reported globally up to a 2,600 cubic kilometers per year ($km^3\ yr^{-1}$) increase in vapor flux because of irrigation, and northern India has been experiencing one of the most intense levels of vapor flux (500 millimeters per year ($mm\ yr^{-1}$)) because of irrigation (Gordon and others, 2005, figs. 3 and 4). Sen Roy and others (2007) also examined the impacts of irrigated agriculture on observed dry season temperature in northern India, analyzed observed temperature data, and supported the findings with regional modeling. This study found up to 0.34°C cooling of growing (dry) season maximum temperatures after irrigation was adopted. Up to a 0.53°C decrease of maximum temperature was reported for individual growing season months. This study reported increased latent energy flux and regional moisture flux because of irrigation. It was concluded that these lowered maximum temperatures were forced by increased evaporative cooling and decreased Bowen ratio (that is ratio of sensible to latent heat fluxes) associated with irrigation. The study also reported statistically significant long-term negative temperature trends.

In a modeling study, Puma and Cook (2010) found notable regional cooling effects over southern and eastern Asia early in the 20th century. These impacts were intensified and became significant across the middle-latitude croplands during the mid-20th century. The results suggest winter season warming of parts of North America because of increased irrigation. Puma and Cook (2010) concluded that irrigation will play an important role in the future global climate.

Biggs and others (2008) also found lowered maximum temperature, increased latent and decreased sensible energy fluxes because of increased irrigation over the Krishna River Basin in southern India. Recent work by Douglas and others (2006, 2009) also reported decreased latent and increased sensible energy fluxes in India because of adoption of irrigated agriculture. Increased regional moisture content, lowered PBL height, and increased convectively available potential energy (CAPE) also were reported by Douglas and others (2006, 2009). Lowering of temperature in Australia because of irrigated agriculture (thus an increase in latent energy flux) also is reported by Geerts (2002).

2.2.1.3 Precipitation

In addition to temperature, a number of studies reported increases in precipitation because of the introduction of irrigated agriculture. For example, Barnston and Schickedanz (1984) noted an increase in precipitation because of irrigation over the Ogallala aquifer region and Southern Great Plains. The observational data-based study found that precipitation increased when low-level convergence and uplift were present within a synoptic setting and allows low-level moisture to rise and reach the base of clouds. As a result, the impact of irrigation was greater for prevailing rain-producing conditions that typically allowed such uplift. In recent followup work, DeAngelis (2010) reported up to a 15–30-percent increase in July precipitation over the easternmost part of the Ogallala aquifer. The influence of irrigation and these changes were extended as far east as Indiana.

In regard to large-scale LULC, the Corn Belt depicts a seasonally invariant and relatively undifferentiated structure to the land cover. However, satellite-derived indices of vegetation activity, such as the Normalized Difference Vegetation Index (NDVI), reveal strong heterogeneity in the early warm season that is related to soil moisture content, crop stage of development (phenology), and vegetation type. The presence of long yet relatively narrow boundaries between extensive croplands and remnant forest serves to enhance convective cloudiness and rainfall during the warm season, supporting Anthes' (1984) model results that widely-spaced rows of trees may initiate or intensify deep convection at least in the subtropics. This land-surface-convection coupling is most evident when the synoptic-scale atmospheric circulation favors weak winds in the free atmosphere. It is hypothesized that the large height differences between trees and crops can loft air parcels above the condensation level when humidity is high in the layer of air closest to the ground. The vegetation-boundary association with precipitation may be seasonally varying and tied to phenology differences between the croplands and surrounding forest areas. Early in the warm season, longer and wider boundaries help promote more precipitation, while later in the summer, shorter and narrower boundaries have associated high precipitation amounts. It is hypothesized that the seasonal switch in maximum NDVI between the forests and croplands changes the "pooling" of moisture along the vegetation boundaries.

In a modeling study, Costa and others (2007) found that replacement of forests in Amazonia with soybean farmland significantly reduced precipitation compared to replacement with pastureland. This study suggests that lowering of albedo and latent heat flux resulted in a negative feedback suppressing convection, cloudiness, and hence, precipitation. In followup research, Nair and others (2011) and Junkermann and others (2009) reported that replacement of native vegetation with agriculture also reduced rainfall. It was suggested that among other influences, changes in surface roughness play a critical role. Recently, Sen Roy and others (2011) analyzed observed data and found increased premonsoon rainfall over irrigated areas of northern India. Supporting modeling work found increases in latent heat flux, equivalent temperature, and lowered PBL that may have played important roles in this increase. Lee and others (2009) noted that LULC change in India has impacted early summer monsoon rainfall. and report increased anomalies in NDVI for March, April, and May because of increases in irrigation. In addition, significant decreases in July rainfall in central and southern India were linked to the greening vegetation as measured by an NDVI increase in the preceding March, April, and May period. Saeed and others (2009) also reported modification of premonsoon and monsoon precipitation in various regions of the Indian subcontinent because of irrigation.

2.2.2 Land use and Land Cover Because of Urbanization and Changes in Weather and Climate

2.2.2.1 Temperature

Urbanization provides examples of intense localized changes in LULC and its impact on weather and climate. One of the most well-known impacts of urbanization is the urban heat island (UHI) (Landsberg, 1970; Arnfield, 2003; Souch and Grimmond, 2006; Yow, 2007). UHI can be characterized by higher air temperatures compared to the surrounding rural area. It is found that UHI can be influenced by time of day, seasons, latitude, climate regime, and surrounding land cover (Arnfield, 2003; Zhou and Shepherd, 2010. In a recent study, Imhoff and others (2010) found that ecological context influences summer-season daytime amplitude of the UHI.

2.2.2.2 Precipitation

It has also been found, particularly in recent literature, that urbanization impacts convective precipitation (Shepherd and others, 2002, 2010; Shepherd, 2006; Niyogi and others, 2006; Kauffman and others, 2007; van den Heever and Cotton, 2007; Mote and others, 2007; Lei and others, 2008; Rose and others, 2008; Stallins and Rose, 2008; Trusilova and others, 2008; Hand and Shepherd, 2009; Shem and Shepherd, 2009; Kishtawal and others, 2010; Niyogi and others, 2010; Niyogi and others, 2011; Mitra and others, 2011). The impacts include changes in location of precipitation compared to pre-urbanization, enhancement in actual amounts of precipitation, and development of preferential location of precipitation as a result of urbanization. Overall, these studies suggest that urbanization modifies surface physical characteristics and energy partitioning and leads to changes in local atmospheric circulation and convection that eventually impact precipitation.

2.3 Land Use and Land Change and Biogeochemical Cycles

Land-cover changes can influence climate change through biogeochemical cycles, especially the carbon (C) and nitrogen (N) cycles. Several primary greenhouse gases that influence climate change are directly involved in ecological systems. They have the following influence on the radiative forcing that causes global warming: 50 percent by carbon dioxide (CO_2), 19 percent by methane (CH_4), and 4 percent by nitrous oxide (N_2O) based on 1990 concentrations, excluding water vapor, and accounting for the residual time of each gas in the atmosphere (Houghton and others, 1990). There is considerable interest in offsetting some of the increase in greenhouse gases from burning fossil fuels by reducing the emissions of these gases from ecological, agricultural, and forest systems, or increasing the uptake of CO_2 by these systems.

For the time scales of interest in climate change mitigation and adaptation (tens of years to centuries), the significant pools of carbon of interest are the atmosphere, vegetation, soils, and the oceans. At the current levels, the atmosphere has about 750 petagrams of carbon (Pg C), the soils have about 1,500 Pg C in the top 1 m or up to 2,300 Pg C in the top 3 m, and the terrestrial vegetation has 850 Pg C (Eglin and others, 2010). The oceans have a very large pool and are important in the time scales of interest for absorbing about 1/3 of global CO_2 emissions.

Approximately 120 Pg C is taken from the atmosphere each year by terrestrial ecosystems through photosynthesis of which about half is quickly (seconds to days) released back to the atmosphere as CO_2 by plant respiration, and the other half is released slowly by respiration or decomposition of plant and animal materials. A small amount of the annual gross primary production is incorporated into soil organic matter, which is often modeled as having three pools with turnover times of approximately 1.5 year, 25 years, and 1,000 or more years (Parton and others, 1987). In modeling grassland soils using a simplified version of the Century model, Parton and others (2001) indicated that approximately 48 percent of the soil organic carbon (SOC) was in the slow pool (10- to 30-year turnover time), approximately 50 percent of the SOC was in the passive pool (1,000- to 3,000-year turnover time), and about 2 percent was in the microbial pool (active decomposition).

Most of the carbon released from soils is as CO_2, with a smaller amount released as CH_4. Both of these are greenhouse gases, and although there are fewer molecules released as CH_4, the heating influence of each CH_4 molecule on the radiation balance of the earth is about 25 times more than the influence of each CO_2 molecule (McGuire and others, 2010; citing Intergovernmental Panel on Climate Change, 2007). Release of CH_4 from soils generally is associated with wetland soils or rice cultivation. Releases of N_2O generally are associated with agricultural activities such as nitrogen fertilization. N_2O is even stronger per molecule than CH_4 in its warming influence on the Earth's radiation balance.

Because the soil carbon pool is large in comparison to the biospheric and atmospheric pools, a small change in the soil pool can result in a larger proportional change in the atmospheric pool. Therefore, changes in soil processes resulting from LULC change may have an amplified impact on the atmospheric pool, where increasing levels of greenhouse gases are major drivers for climate change and global warming. If climate change stimulates a net release of greenhouse gases from soils, then it represents a positive feedback contributing to additional climate change.

Land-cover change (for example, forest or grassland to agriculture) and land-use and management practices (for example, crop choice, fertilization rates, type of tillage, and use of irrigation) can change the soil conditions that affect the rates of release of SOC and N_2O. The release rates also are influenced by the type of soil (for example, soil texture, water holding capacity, permeability, organic matter content, drainage, and erosion rate) and the seasonal patterns of soil moisture and temperature. The patterns of soil moisture and temperature are expected to change with changes in climate. Land-management practices also may be adjusted with climate change (for example, influences of drought or floods on agriculture) and by human attempts to mitigate the drivers of climate change (for example, making biofuels with feedstocks from croplands, grasslands, or forests to use as alternatives to fossil fuels).

Guo and Gifford (2002) provide a global review of the effects of land-cover change on soil carbon stocks. Meta analysis of 74 publications by Guo and Gifford gives the following results in terms of percentage change in SOC stocks:

-10 percent	pasture to plantation (for example, plant trees),
-13 percent	native forest to plantation,
-42 percent	native forest to crop,
-59 percent	pasture to crop,
+8 percent	native forest to pasture,
+19 percent	crop to pasture,
+18 percent	crop to plantation, and
+53 percent	crop to secondary forest.

In general, conversion of native forest or grassland to cropland results in a large loss of soil carbon (presumably to the atmosphere and within the first few years after conversion). Conversely, converting cropland into forest or grassland can store atmospheric carbon in the soil, but it may take 70 or more years and the new "equilibrium" may be at a lower level of SOC than in an ecosystem that was never disturbed.

Increases in carbon stored in vegetation or soil are often referred to as "carbon sequestration" because they remove carbon from the atmosphere for some period of time. Some of the increases in storage in vegetation are attributed to the "CO_2 fertilization effect" in which higher levels of CO_2 in the atmosphere cause plants to grow faster (CO_2 is an input to photosynthesis, so more CO_2 may lead to more photosynthesis).

Eglin and others (2010) cite the global soil carbon stocks reported by Jobbagy and Jackson (2000) and assess the global vulnerability of SOC to climate and land-use changes. They note that with increasing CO_2 in the atmosphere the size of the terrestrial sink (annual storage) also has been increasing from 1.8 PgC in the 1980s to 2.6 PgC in the 1990s and 3 PgC from 2000 to 2008. They also note that the land biosphere carbon sink absorbs about 30 percent of the anthropogenic emissions from fossil fuel combustion and deforestation every year (Le Quere and others, 2009; Canadell and others, 2007; Intergovernmental Panel on Climate Change, 2007). Although soil contains three times more C than vegetation, it is a two times smaller sink than living biomass (Reichstein, 2008 as cited by Eglin and others, 2010) because (1) when part of the biomass is harvested, the soil does not receive carbon from the harvested productivity, and (2) increased primary productivity adds C into biomass pools where it could be stored for decades as wood in forests and is not quickly passed into the soil. The carbon in soils may become more vulnerable to decomposition in a warmer world, and Reichstein (2008) estimates the potential loss of soil C under the future warming forecasted by climate models is six times larger than the current soil C sink.

Although there has been a lot of research on the potential for changes in land use and management to mitigate the increases of greenhouse gases in the atmosphere, there may be a limited potential for this. In one of the most complete analyses at the national level, Smith and others (2010) assessed how agricultural land-use changes influence SOC and greenhouse gas emissions in Great Britain. Transitions were considered between cropland, temporary grassland (less than 5 years under grass), permanent grass (more than 5 years under grass), and forest. It was shown that reversion to historical land-use patterns as present in 1930 could result in greenhouse gas emission reductions of up to 11 teragrams of carbon dioxide equivalent per year (Tg CO_2-equivalent yr^{-1}), mostly because of an increased permanent grassland area. By contrast, cultivation of 20 percent of the permanent grassland area (as of 2004) for crop production could result in greenhouse gas emission increases of up to 14 Tg CO_2-equivalent yr^{-1}. It was concluded that changes between agricultural land uses in Great Britain is likely to be a limited option for greenhouse gas mitigation, mostly because it is unlikely that existing cropland would be transferred into grassland. In terms of total United Kingdom greenhouse gas emissions, even the most extreme feasible land-use change scenarios would offset only about 2 percent of current national greenhouse gas emissions (Smith and others, 2010).

The LandCarbon project of the USGS is evaluating carbon storage, carbon fluxes, and N_2O fluxes in all forests, grasslands/shrublands, agricultural lands, wetlands and freshwater aquatic systems in the United States. The assessment uses remote sensing, statistical methods, and simulation models. The study includes a baseline time period (generally in the first half of the 2010s) and future projections from the baseline period to 2050. An initial report covers the Great Plains region (Zhu and others, 2011). Among the conclusions is an estimated annual carbon sequestration rate for the region between 20 and 99 TgC yr^{-1} from 2001 to 2005, and an estimate for the mean annual sequestration rate of 48 to 66 TgC between baseline and 2050. The study methods include forecasting LULC change using Intergovernmental Panel on Climate Change (IPCC) scenarios, driven by the demand for agricultural commodities (including biofuels). The scenarios suggest a significant expansion of agricultural land (1.4 to 9.2 percent of the total area by 2050, depending on which scenario is used in the calculation).

2.4 Combined Impacts of Land-Use and Land-Cover Change and Climate Change

A limited number of studies have been conducted on this topic over the last decade. In one of the earlier studies, Zhao and others (2001) showed LULC change did not get diminished with increasing

CO_2 impacts. This study also identified remote responses because of combined changes in LULC and CO_2 levels. One of the four regions that showed remote response is North America and includes the United States. The remote responses include changes in temperature, which were largely influenced by the changes in circulations. In another modeling study, Pongratz and others (2010) reported that LULC change will lead to, on the average, small global cooling (-0.03 Kelvin (K)) while warming caused by CO_2 will be greater (0.16–0.18 K). In another detailed model-based study Brovkin and others (2004) found net cooling over northern latitudes including most of the North America and the United States because of the combined impact of LULC change and CO_2. However, a relative decline in the role of LULC change compared to CO_2 was noted. Additional studies conducted in Australia also show LULC change under increasing CO_2 may mitigate impacts of the latter (Narisma and Pitman, 2004, 2006). Additional research needs to be conducted to better understand the combined impacts.

3.0 Recent Land-Use and Land-Cover and Climate Case Studies

Case studies selected from current literature (2009 through 2011) demonstrate historic, present, and potential future climate influences on LULC. The case studies were synthesized and organized into the following LULC/environment effect categories: water resources, transportation and energy supply and use, agriculture, ecosystems, human health. The individual case study summaries can be found in appendix I.

3.1 Climate Influences on Land-Use and Land-Cover and Water Resources

Case studies addressing climate influences on LULC and water resources highlight five relevant topics: baseflow, drought and flood, water supply, water quality, and wetlands. Collectively, these case studies show the clear link between climate and LULC and the current state of water resources within the United States.

The case studies on baseflow and flood and drought identified LULC and climate as two main drivers of altered baseflow (Kochendorfer and Hubbart 2010; Price 2011; Price and others 2011; Viger and others 2011). Baseflow, or in hydrologic terms, streamflow results from precipitation that infiltrates into the soil and eventually moves through the soil to the stream channel (also called low flow, National Oceanic and Atmospheric Administration, 2012). Recent changes in baseflow caused by LULC changes alter the natural habitat within watersheds. When natural habitat within a watershed is disturbed, the hydrologic functions such as infiltration and recharge are compromised leading to exacerbated floods in wet years and droughts in dry years. Tomer and Schilling (2009) present the difficulty of separating the climate influences from weather influences and land-use change on watershed hydrology and were able to separate interactions for a 25-year, small watershed experiment in Iowa. If applied regionally, the conclusions suggest that climate change has increased discharge from Midwest watersheds, especially since the 1970s.

Praskievicz and Chang (2009) and Polebitski and others (2011) have studied climate change and urban development and the impacts on water supply and demand. These studies demonstrated that climate change will increase competition between land uses for water resources, particularly in areas of urban development and arid and mountainous regions of the United Sates. Ford and others (2011) looked at the impacts of LULC on forested watersheds and found that unmanaged watersheds had lower streamflow and a more stable water supply.

Water quality can be impacted by several sectors (agriculture and energy supply and use), which ultimately compromises other sectors (water resources, human health, and society). Water-quality case studies suggest a positive relation between increased fertilizer usage and increased phosphorus, nitrogen, and nitrate runoff into streams or groundwater (Banner and others 2009; Burow and others 2010; Broussard and Turner, 2009). Another study found that land-cover changes to natural vegetation reduced the watershed's ability to retain nitrogen, and thus led to runoff and eutrophication (Kinney and Valiela, 2011).

LULC change involving wetlands affects both water resources and ecosystems. Wetland functions are sensitive to agricultural practices and climate. Adaptations in agricultural management may help ameliorate climate change by retaining carbon sinks and water levels (Johnson and others, 2010, Kumar and others, 2010 and Voldseth and others, 2009).

3.2 Climate and Land-Use and Land-Cover Influences on Transportation and Energy Supply and Use

The impacts of transportation and energy supply and use sectors are similar. For example, transportation is essential to providing energy supply and use. DellaSala and others (2011), Duniway and others (2010), and McDonald and others (2009) have assessed the impact that transportation and energy activities have on the landscape. In Colorado, development of roads within undeveloped watersheds could potentially enable short-term energy supply (production of oil and gas); however, the long-term impact would reduce water supply and quality (DellaSala and others, 2011). Continued development of oil, gas, and renewable energy sources would require service roads to be developed in rangelands (Duniway and others, 2010). Although climate influences were not specifically mentioned in DellaSala and others (2011), Duniway and others (2010), and Dale and others (2011) suggested that analysis of landscape patterns could be used to enhance the understanding of interactions among land use, energy, and climate change. McDonald and others (2009) found that even if there is no cap-and-trade bill to reduce emissions, 206,000 km^2 of land will be affected by energy production by 2030. Ethanol has been used to help reduce emissions; however , Solomon (2010) found that corn-based ethanol was unsustainable and had significant environmental costs. Interactions between transportation and energy have significant influences on LULC and climate and are expected to increase as the demand for energy also increases.

3.3 Climate Influences on Land Use and Land Cover and Agriculture

Agricultural production and land-use practices are influenced by climate, land, and water resources. Much of the world depends upon agriculture in the United States to supply many goods and services. As the world's population continues to grow, so does the demand for increased agricultural production (Burow and others, 2010). Associated impacts of this demand include increased fertilizer usage, which results in increased concentrations of phosphorus and nitrogen loading in watersheds (Banner and others, 2009; Broussard and Turner, 2009; Burow and others, 2010). Land-use conversion to cropland, cropland sustainability, and increased growing season temperatures were identified as agricultural topics that were impacted by climate and land-use change.

3.4 Climate Influences on Land Use and Land Cover and Ecosystems

Climate, LULC, and ecosystem issues include greenhouse gas inventory and biogenic emissions, vegetation displacement, and wildland fires. The United Nations Framework Convention on Climate Change (UNFCCC) requires all qualifying nations to report the status of their greenhouse gas (GHG) emissions and sinks annually (United National Framework Convention on Climate Change, 2012a, 2012b). Zheng and others (2011) estimated the effects of major forest disturbances and net growth on carbon sequestration in the conterminous United States, based on the terminology and requirements for reporting to the UNFCCC for national GHG inventories. There are two types of emissions defined by the Environmental Protection Agency—anthropogenic and biogenic emissions (U.S. Environmental Protection Agency, 2012). Anthropogenic emissions are produced as a result of human activity that releases CO_2 emissions into the atmosphere. One of the largest sources of anthropogenic CO_2 emissions is the combustion of fossil fuels or fossil fuel-based products to produce electricity (U.S. Environmental Protection Agency, 2012). Biogenic emissions, in contrast, result from natural biological processes, such as the decomposition or combustion of vegetative matter. Biogenic CO_2 emissions are part of a closed carbon loop. Biogenic CO_2 emissions are balanced by the natural uptake of CO_2 by growing vegetation, resulting in a net zero contribution of CO_2 emissions to the atmosphere. Examples of biogenic emission sources include burning vegetation (biomass) to produce electricity or using plant-based biofuels for transport (U.S. Environmental Protection Agency, 2012). Studies have been done to quantify the effects of LULC on anthropogenic emissions; however, few studies have looked at the effects of LULC on biogenic emissions. Chen and others (2009) used a model to study LULC effects on future biogenic emissions within the United States.

Projected climate change will have an impact on the distribution of native vegetation. It has been projected that the natural climate gradient of forest ecosystems will shift to higher elevations, thus displacing certain tree species (Tang and Beckage, 2010). Tang and Beckage (2010) studied the three major forest types in New England and modeled each forest's distribution in response to climate change. In other ecosystems, climate change would impact native vegetation surviving and their ability to compete with invasive species. Bradley (2010) studied the risk of sagebrush loss associated with climate and land-use change. Barret and Gray (2011) presented the potential benefits of extending the Forest Inventory and Analysis (FIA) approach to the boreal forest region of Alaska and suggested that FIA monitoring could aid conservation decisions by providing information on the abundance and rarity of vascular plants, invasive species, biomass and carbon content of vegetation, shifting vegetation species distribution, disturbance frequency, type, impact, and wildlife habitat characteristics.

Climate change is already altering natural fire regimes within the United States (Liu and others, 2010; Westerling and others, 2011). Increased duration of fire seasons accompanied by drought conditions leads to increased wildfire potential. However, climate-driven changes in regional fire regimes are not well understood (Westerling and others, 2011). Liu and others (2010) and Westerling and others (2011) measured fire potential and predicted future fire trends in response to climate change. Westerling and others (2011) also mentioned that increased fire frequency would transform fire regimes and change vegetation types.

3.5 Climate Influences on Land Use and Land Cover and Human Health

Projected climate change could/may have an impact on human health with LULC implications. Two health related topics identified in the literature as being linked to LULC and climate influences were West Nile virus and extreme heat events. Two studies examined the climate and land-use drivers of the Northern Great Plains because of the number of cases found within this region. Wimberly and others (2008) assessed environmental drivers of West Nile virus, and Chuang and others (2011) studied land cover and climatic variability of the host mosquitos. Extreme heat events are becoming a concern in large metropolitan areas because of increasing frequency. Gershunov and others (2009) studied the 2006 heat wave over California and Nevada in context to the region's climate over the last 60 years. Associations between metropolitan region and the frequency of extreme heat events over a five-decade period were examined using a sprawl index (Stone and others, 2010).

3.6 Climate Influences on Land Use and Land Cover and Society

Climate and LULC can combine to cause negative feedbacks on society. For example, urbanization can affect regional weather and climate and cause land fragmentation. In the Southeastern United States, Nagy and Lockaby (2010) tested the hypothesis that a negative feedback would exist between environmental impacts and the rate and patterns of development. Niyogi and others (2011) studied the impact an urban region would have on altering the intensity, structure, and composition of an approaching thunderstorm. In the Southwestern United States, York and others (2011) investigated rapid urbanization to determine drivers that caused land fragmentation.

4.0 A Regional and Sectoral Perspective of Climate and Land-Use and Land-Cover Interactions

Climate and land-system interactions are as unique as the regional and sectoral driving forces behind them. This section presents an integrated framework for assessing climate change and variability and land-change feedbacks, vulnerability, and resilience. The framework presented in figure 4 (McMahon and others, 2005) acknowledges that variability and change in land systems should be identified, which is what this report presented in the previous section on case study analysis. The framework also acknowledges that consequences and implications of land change, such as vegetation response to drought, should be identified. To demonstrate this part of the framework, this section discusses regional and sectoral responses to drought from 2009 through 2011 by presenting results from the USGS's Vegetation Drought Response Index (VegDRI) project (Brown and others, 2008). Lastly, the framework presents the need to understand human-environmental system vulnerability to climate and land change by investigating regional responses to change through adaptation strategies. While not conclusive, this section reviews possible adaptation strategies for the agricultural, forest, water, and wetland land-cover sectors.

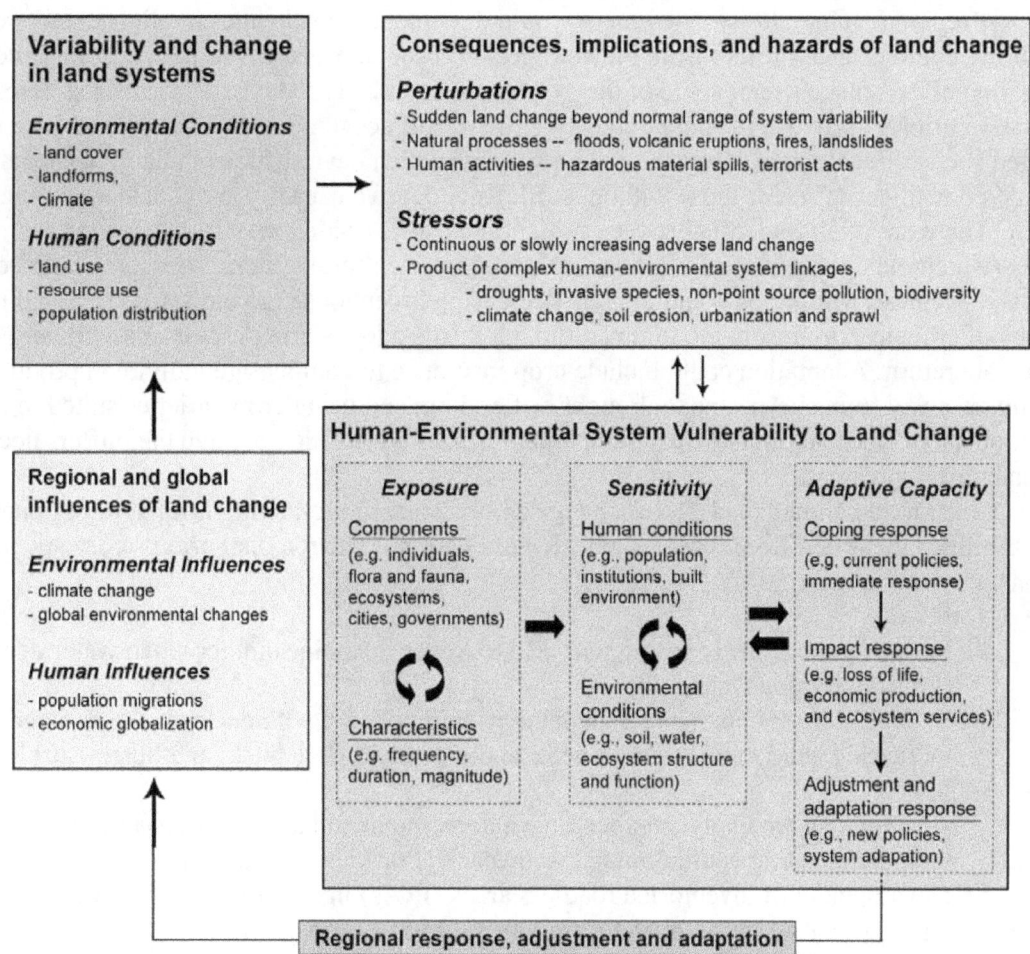

Figure 4. A coupled framework for assessing land change, vulnerability, and adaptability (McMahon and others, 2005; modified from Turner and others, 2003).

4.1 Identification of Vulnerable Regions and Sectors

The vulnerability of LULC is based on the degree to which a coupled human-environmental system is susceptible to reduced function, goods, and services because of adverse effects of climate change, including climate variability and extremes. There are multiple feedbacks in vulnerable systems that are functions of their exposure, sensitivity, and adaptive capacity to the character, magnitude, and rate of climate change. In the figure 4 (yellow box), environmental and human influences are shown that can contribute to change in the environmental and human conditions in land systems (green box). There are many processes (blue box) operating on short-term (perturbations) or longer-term (stressors) time scales that influence how flora, fauna, individuals, and institutions use the land to exist and produce things (pink box). The capability of people and institutions to adapt may depend on the extent to which they have maintained an adaptive capacity (for example, set aside resources for responding to hazards). Some of the adaptive responses may be to change the ways in which humans influence the land system (orange box and yellow box). The boxes and arrows in this figure are conceptual because in a real system there are no clear boundaries between the primary human or ecological processes and the ways in which they mitigate or adapt to stress.

23

A regional vulnerability assessment would document how changes in climate and weather variability could change the exposure of ecological and social processes to perturbations and stressors. Mitigation efforts make use of some of the system's resources to reduce the harmful stresses, and adaptation efforts make use of some of the system's resources to reduce the harm or disruption to affected social or ecological subsystems. The systems which have sufficient adaptive capacity are considered resilient, whereas those without sufficient adaptive capacity are considered vulnerable.

There are often tradeoffs between mitigation efforts (which may be long term) and adaptation efforts (which may be short term). For example, in an agricultural system, stressors could be droughts or excessive wetness. Mitigation could include decreasing greenhouse gas emissions to avoid increasing the rate of climate extremes in the future, although this is a long-term process and will not guarantee an immediate return. Adaptation could include crop insurance to compensate individual producers, moving agriculture away from sites prone to drought or flooding, and using crop varieties suited to the changed conditions. Systems that operate near 100 percent capacity often do not have the buffers needed to withstand new stressors.

Considering climate and LULC change as stressors, the following is a list of regional vulnerabilities identified from recent scientific publications. The regional groups correspond to the NCA regions.

- Northwest
  - Earlier peak runoff in spring will lead to lower flows in summer when water demand is high (Praskievicz and Chang, 2009).
  - In Seattle, it was predicted that water pricing policy would decrease the demand; however, increasing temperature would increase the demand (Polebitski and others, 2011).
- Southwest
  - Arid regions are likely to experience a decrease in annual runoff because of increased temperatures that could diminish snowpack (Praskievicz and Chang, 2009).
  - Development of inventoried roadless areas (IRAs) in Colorado could have long-term impacts on the water supply (DellaSala and others, 2011).
  - Continued development of renewable energy sources, including wind and solar, will require service road development in rangelands (Duniway, 2010).
  - The risk of sagebrush loss associated with land use and cheatgrass invasion was highest in portions of Nevada already dominated by cheatgrass and near lands used for agriculture (Bradley, 2010).
  - Continued warming could transform Greater Yellowstone ecosystems fire regime by the mid-21st century with a likelihood that forested ecosystems would be converted to nonforest vegetation and thus alter the fire return interval (Westerling and others, 2011.
  - Recent increases in California and Nevada heat waves are consistent with regional signs of global warming (Gershunov and others, 2009).
  - Five drivers of land fragmentation patterns were identified including: water provisioning, urban population dynamics, transportation, topography, and institutional factors. Water was identified as the key variable in understanding land change in the Southwest United States because all five study sites had major rivers dammed for storage or prevention of flooding (York and others, 2011).
- Great Plains and Midwest
  - Rural land-use change has played a significant role in streamflow alteration (Kochendorfer and Hubbart, 2010).

- o Watersheds that have been under agricultural land use for an extended period of time have increased baseflows because of improved cropping and tillage practices (Price, 2011).
- o Climate change is likely to exacerbate conflicts over competing land uses for water resources, such as urban and agriculture (Praskievicz and Chang, 2009).
- o Fertilizer usage has increased in order to increase crop production (Banner and others, 2009; Broussard and Turner, 2009; Burow and others, 2010).
- o Increased fertilizer usage has increased phosphorus, nitrogen, and nitrate runoff into streams and groundwater (Banner and others, 2009; Burow and others, 2010; Broussard and Turner, 2009).
- o Corn, soybeans, and cotton increased yield threshold with increased temperatures of 29oC, 30oC, and 32oC, respectively (Schlenker and Roberts, 2009).
- o Sensitivity of wetlands to warming temperatures within the Prairie Pothole Region (Johnson and others, 2010).
- o Agricultural drainage of Midwest wetlands have changed the carbon balance from carbon sinks to carbon sources and have affected the roles wetlands play in climatic feedbacks (Kumar and others, 2010).
- o Adaptations of agricultural practices can offset climate change effects on wetlands (Voldseth and others, 2009).
- o Temperate grasslands were identified as one of two most impacted habitats by future energy development (McDonald and others, 2009).
- o Increased market prices of corn and soybeans throughout the Midwest will contribute to a considerable decrease of lands enrolled in conservation programs (Gallant and others, 2011).
- o Grassland conversion to cropland was found to be driven by increased economic returns and the probability of grassland conversion to cropland will increase making grassland conservation more important (Rashford and others, 2011).
- o Incidences of West Nile virus are higher in the northern Great Plains than in other regions of the United States because the environmental conditions create a favorable ecological niche for Culex tarslis, a known vector of the West Nile virus (Wimberly and others, 2008).
- o Temporal and spatial distributions of two important mosquito species in South Dakota and their relations with land cover and weather will help to enhance the efficiency of vector control efforts and disease prevention (Chuang and others, 2011).
- o Thunderstorms were analyzed to capture varying storm events around the Indianapolis urban region and peripheral rural counties; it was found that 60 percent of storms changed structure over the Indianapolis area as compared to 25 percent over the rural regions (Niyogi and others, 2011).
- Southeast
  - o High-forest covered watersheds had lower baseflows, compared with areas of low- forest covered watersheds. Infiltration and recharge under undisturbed land cover is important in order to sustain low flows (Price, 2011).
  - o Urbanization and climate change could potentially affect the freshwater resources of the Flint River System because of decreased streamflow caused by a decrease in surface-runoff and groundwater components (Viger and others, 2011).
  - o Forest management of forested watersheds could potentially help mitigate extreme annual precipitation by decreasing flood risks because of pine forests ability to increase soil water storage (Ford and others, 2011).

- o Environmental conditions had little influence on rates and patterns of development in the west Georgia Piedmont region because of increasing land values, market concentrations, and road accessibility (Nagy and Lockaby, 2010).
- Northeast
  - o Land cover shifts away from natural vegetation will increase nitrogen loading and lead to eutrophication of Great South Bay, New York (Kinney and Valiela, 2011).
  - o Temperate deciduous forests along with temperate grasslands (Midwest) were identified as the two most impacted habitats by future energy development (McDonald and others, 2009).
  - o Climate warming in the 21st century is likely to cause the extensive loss of boreal conifer forests, reduce the extent of northern hardwood deciduous forests, and result in large increases of mixed oak-hickory forest (Tang and Beckage, 2010).

4.2 Ongoing Land-Use Adaptations to Climate Change

The importance of developing land-management options for climate change is anchored by an understanding of the regional variability of human-environmental interactions across the United States. These interactions include but are not limited to the rates, driving forces, and implications of land change. As science-management partnerships are established, it becomes increasingly important to identify how specific land-use practices are responding to climate change and what specific adaptations should be considered for resource conservation. General land-cover classes that have an interface with the human environment and play a large role in resource conservation and management in the United States include agriculture, forest, water, and wetlands. The following sections summarize land-use adaptations by sector that were reported in recent publications.

4.2.1 Agriculture

In addition to adapting to a changing climate, the agricultural sector will also face sustainment challenges associated with an increasing population. Agricultural management decisions may want to consider various land uses such as cropping systems and livestock in relation to climate variables such as warming and fluctuations in precipitation.

Agricultural land managers should be aware of feedbacks associated with sectoral changes. For example, Voldseth and others (2009) examined the effects of climate and watershed cover on wetland water levels in eastern South Dakota to investigate if adaptation of agricultural land-use practices ameliorates effects of climate change on prairie wetlands, and how much climate change can be offset or absorbed through land-use management. This study concluded that cultivated crops and managed grasslands returned water levels that were equal or greater than unmanaged grassland under historical climate for the $2°C$ temperature increase and the $2°C$ and 10 percent precipitation increase. Managed grasslands returned water levels that were equal or greater than unmanaged grasslands under historical climate for the $4°C$ temperature increase and the $2°C$ and 10 percent precipitation increase.

There are also economic impacts to consider when implementing climate adaptation strategies for the agricultural sector. For example, for livestock systems, Howden and others (2011) suggest that agricultural land-use adaptations to climate change include modifying grazing timing and paying special attention to continuously match stock rates with pasture production. The authors go on to discuss impacts on livestock agriculture by pointing out that more heat-tolerant livestock breeds often have lower levels of productivity. See table 3 for a summary of agriculture adaptations.

Table 3. Summary of agricultural land-use and cover adaptations with corresponding literature citations.

Land Cover	Land Use	Climate Variable	Adaptation	Reference
Agriculture	Cropping Systems	General Change	Altering varieties and species	Howden and others, 2007; Delgado and others, 2011; Hallegatte, 2009; Yadav and others, 2011; Easterling, 1996; USGCRP, 2000;Voldseth and others, 2009
			Altering fertilizer rates to maintain quality consistent with prevailing climate	Howden and others, 2007; USGCRP, 2000
			Altering amounts and timing of irrigation	Howden and others, 2007;Voldseth and others, 2009
			Alter timing or location of cropping activities (to include crop rotation and cover crop use)	Howden and others, 2007; Delgado and others, 2011; Easterling, 1996; USGCRP, 2000
			Increase soil C sequestration to improve soil functions	Delgado and others, 2011
			Increase N-use efficiencies for cropping systems	Delgado and others, 2011
			Diversify income by integrating other faming activities such as raising livestock	Howden and others, 2007
		Rainfall decrease	Adapt technologies to conserve soil moisture (because of crop reside retention)	Howden and others, 2007; Yadav and others, 2011
			Increase irrigation efficiency	Delgado and others, 2011; Easterling, 1996; USGCRP, 2000
		Rainfall increase	Manage water to prevent water logging, erosion, and nutrient leaching	Howden and others, 2007;
	Livestock systems	General Change	Additional care to match stock rates with pasture production	Howden and others, 2007;
			Altered rotation of pastures	Howden and others, 2007;
			Modify times of grazing	Howden and others, 2007;
			Alter forage and species/breeds	Howden and others, 2007;
			Altered integration within mixed livestock/crop systems	Howden and others, 2007;
			Reducing CH_4 emissions from ruminants with feeding management, use of edible oils, and possible vaccinations	Delgado and others, 2011
			Use of supplementary feed and concentrates	Howden and others, 2007

Land Cover	Land Use	Climate Variable	Adaptation	Reference
		Warming	Increased need for management and infrastructure to ameliorate heat-related reductions in productivity, fertility, and increases in mortality	Howden and others, 2007
			Tillage practices that incorporate crop residues would likely combat OM loss due to warming and improve soil quality	USGCRP, 2000
	General agricultural	General change	Use more efficient power sources and renewable energy	Delgado and others, 2011
			Valuing agricultural commodities for their water footprint or environmental traits	Delgado and others, 2011
			Apply precision/target conservation to increase conservation effectiveness across spatial and temporal variability	Delgado and others, 2011
			Adapt farming practices to mitigate impacts of climate change on wetlands as a watershed management option	Voldseth and others, 2009

4.2.2 Forest

Climate change may influence forest-management decisions to identify impacts and decisions that confer resilience and address vulnerability at multiple spatial scales (table 4). Management of forest for timber production may have to consider adaptation strategies such as altering rotation periods and harvesting patterns (Howden and others, 2011), whereas management of national forest land may put a stronger focus on managing ecosystem response to change and fire as both a natural and anthropogenic driver of change (Blate and others, 2009).

Table 4. Summary of forest-related land-use and cover adaptations with corresponding literature citations.

Land Cover	Land Use	Climate Variable	Adaptation	Reference
Forest	**Planted Forests**	General Change	Adapt management intensity	Howden and others, 2007
			Hardwood/softwood species mix	Howden and others, 2007
			Altering harvesting patterns within and between regions	Howden and others, 2007
			Alter rotation periods	Howden and others, 2007
			Salvage dead timber	Howden and others, 2007
			Shift species or areas more productive under new climatic conditions	Howden and others, 2007

Land Cover	Land Use	Climate Variable	Adaptation	Reference
			Make adjustments to altered wood size and quality	Howden and others, 2007
			Adjust fire management systems	Howden and others, 2007
	National Forests	General Change	Increase landscape diversity	Littell and others, 2011
			Maintain biological diversity	Littell and others, 2011
			Implement early detection/rapid response for exotic species and undesirable resource conditions	Littell and others, 2011; Blate and others, 2009
			Treat large-scale disturbance as a management opportunity and integrate it in planning	Littell and others, 2011; Blate and others
			Implement treatments that confer resilience at large spatial scales	Littell and others, 2011;Mawdsley and others, 2009
			Match engineering of infrastructure to expected future conditions	Littell and others, 2011; Blate and others, 2009
			Promote education and awareness about climate change among resource staff and local public	Littell and others, 2011; Blate and others, 2009
			Collaborate with a variety of partners on adaptation strategies and the promotion of ecoregional management	Littell and others, 2011; Blate and others, 2009
			Reduce fuel loads in forests	Blate and others, 2009
			Increase use of wildland fire use	Blate and others, 2009
			Increase efforts to reduce current stress factors	Blate and others, 2009
			Develop silviculture treatments to reduce drought stress	Blate and others, 2009
			Review genetic guidelines for reforestation	Blate and others, 2009
			Provide technical assistance to urban foresters to sustain urban trees	Blate and others, 2009
			Develop corridors for species migration and habitat protection	Blate and others, 2009; Mawdsley and others, 2009
			Evaluate recreational impact on ecosystems under a changing climate	Blate and others, 2009

4.2.3 Water

Water resources include our Nation's surface and groundwater that will face climate-related challenges—irrigation demands supported, fresh drinking water provided to rural and urban communities, and habitat provided for nurturing ecosystems and their services (table 5). Water-resources management also includes mitigating consequences that affect water quality and quantity. For water quality, agricultural and urban runoff are often associated as principal sources of pollution in waterways. But as land cover changes, feedbacks and natural response should be considered when planning adaptation strategies. Kinney and Valiela (2011) used a nitrogen-loading model to: (1) estimate annual nitrogen deliverables to each subwatershed of Great South Bay, new York, (2) partition the contribution of nitrogen through disposal of wastewater, use of fertilizers, and atmospheric deposition on land, (3) estimate nitrogen transport through major land-use categories (natural vegetation, wetlands, agriculture, tuft, impervious surfaces), (4) calculate nitrogen retention within each subwatershed, and (5) model land-derived nitrogen loads from each subwatershed to Great South Bay. Wastewater-derived nitrogen was the dominant source to watershed surfaces followed by atmospheric deposition and fertilizer use. The authors concluded that as land cover shifts away from natural vegetation, eutrophication of Great South Bay will increase.

Table 5. Summary of water resources and land-use and cover adaptations with corresponding literature citations.

Land Cover	Land Use	Climate Variable	Adaptation	Reference
Water	Ecosystems	Warming	Diversify and replicate habitats of special importance	Palmer and others, 2009
			Increase monitoring of populations at high risk	Palmer and others, 2009
			Enhance local monitoring and develop forecasts	Palmer and others, 2009
			Enhance technical assistance at local levels	Palmer and others, 2009
			Enhance protection of rivers (for example to enhance stormwater control and manage riparian corridors)	Palmer and others, 2009
			Use conjunctive groundwater/surface water management (for example drought protection areas, retrofit or remove dams)	Palmer and others, 2009
			Initiate restoration projects (for example to reconnect flood plains)	Palmer and others, 2009
	Hazard mitigation	Warming	Increase off-channel storage basins or wetlands as ways to absorb high flows and maintain water quality	Palmer and others, 2009
	Agricultural	General Change	More efficient water application methods (for example water pulsing) could decrease water needed for agriculture	Ojima and others, 1999

4.2.4 Wetlands

Wetlands adaptations will be dependent on land use and climate feedbacks (table 6). For example, Voldseth and others (2009) suggest that adapting farming practices in response to climate change may influence wetland water levels and land use in the uplands affects the amount and routing of precipitation that reaches wetlands. Therefore, the type of faming practices employed could be used to shift wetland water levels toward drier or wetter conditions in response to climate change.

Table 6. Summary of wetlands and land use and cover adaptations with corresponding literature citations.

Land Cover	Land Use	Climate Variable	Adaptation	Reference
Wetland	Agricultural	General Change	Conversion of high water use crops and unmanaged grassland to managed grassland produced higher wetland levels	Voldseth and others, 2009
			Land use in the upland affects the amount and routing of precipitation that reaches the wetland so the type of farming practices that are employed can be used to shift wetland water levels toward drier or wetter conditions in response to climate change	Voldseth and others, 2009
			Planting dense stands of grass to increase nesting cover in a wetland watershed can dry up wetlands to ensure that wetlands of all permanence types remain in the landscape after reversion to satisfy habitat needs	Voldseth and others, 2009

4.4 LULC and Vegetation Conditions in the United States from 2009–11

Drought is one type of climate extreme that influences land use and impacts human livelihoods. The VegDRI (Vegetation Drought Response Index) data set produced by the USGS has been calculated for every 2-week period since 1989 (National Drought Mitigation Center, 2012). The index incorporates satellite data on vegetation productivity (Normalized Difference Vegetation Index, NDVI), climate data summarized as the Palmer Drought Severity Index (PDSI) and Standard Precipitation Index (SPI), and other biophysical variables such as soil available water capacity, land cover, irrigated lands, and elevation, to estimate drought-related vegetation stress for the conterminous United States (Brown and others, 2008). VegDRI expresses drought categories using the familiar Palmer Drought Severity Index legend. For example, figure 5 shows the VegDRI data for the two-week period ending July 25, 2011, indicating that much of the southern U.S. was in Moderate to Extreme Drought. Consequences of extensive drought may include crop failures, forest die-off, invasive plant species invasion, and water supply depletions.

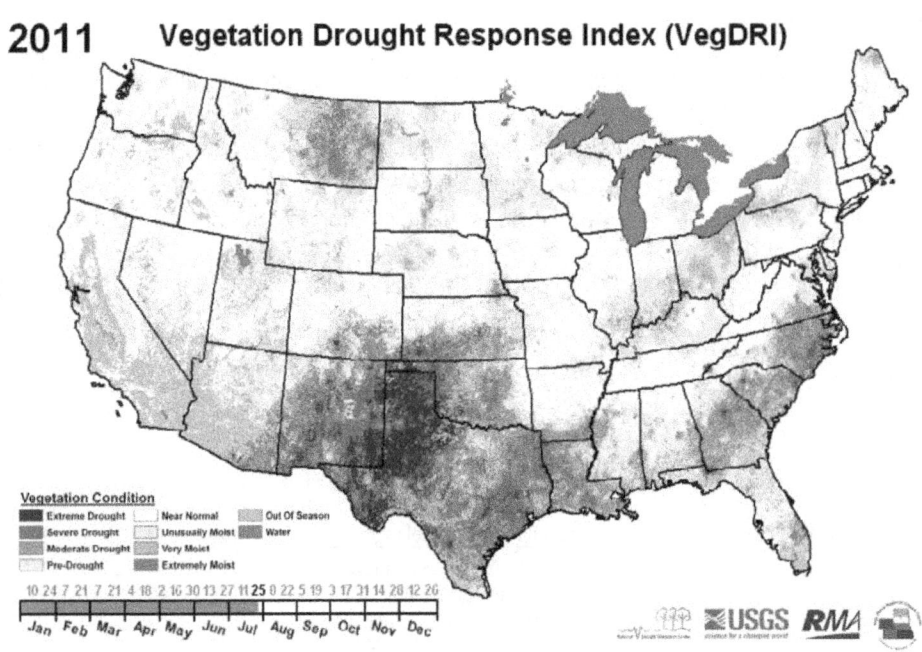

Figure 5. The Vegetation Drought Response Index (VegDRI) map for the 2-week period ending July 23, 2011, shows extensive droughts in the Southern United States (National Drought Mitigation Center, 2012).

      To investigate the interactions of the six conterminous NCA regions in the United States and selected groups of the NLCD land-cover classes, the VegDRI data were summarized for the growing seasons of 2009 through 2011. A growing season summary of VegDRI will embody longer-term drought effects but shorter-term droughts (for example, less than 2 months) may be missed. A summary of the percentage of the area for each land-cover group and region that is in the moderate to extreme drought categories is given in table 7. The average area of land in each of the land-cover categories is given by region in table 8 because there are substantial differences in the area of each of the categories. A more complete view is given in tables 9–14, which in addition to the "drought" category include information on the "normal" and "moist" conditions. In these tables, "Drought" includes the extreme drought, severe drought, and moderate drought categories "Normal" includes the mild drought, normal, and incipient moist spell categories and "Moist" includes the unusually moist spell, very moist spell, and extreme moist spell categories. The analysis was performed using provisional data with a pixel size of 1 square kilometer, so there is more uncertainty when the total areas are relatively small. For example, wetland areas in the Great Plains may be under-represented at this resolution, where many of the wetlands are small. Additionally, while a "developed" category is included in the land cover groups, identifying drought impacts across developed areas is not consistently performed in VegDRI models. The management of developed lands causes confusion in drought signals. The NLCD categories labeled Barren and Shore were excluded from this analysis because the VegDRI was not designed to evaluate these categories (54,957 km^2 nationally).

Table 7. Percentage of land in the National Climate Assessment (NCA) regions and by groups of National Land Cover Database (NLCD) land-cover classes that experienced moderate to extreme drought for the years 2009 to 2011 (original analysis).

Land cover	Year	Northeast	Southeast	Midwest	Great Plains	Northwest	Southwest
Developed	2009	6.7	11.2	6.8	17.9	24.4	22.7
Developed	2010	4.2	8.8	4.8	4.3	6.4	8.4
Developed	2011	2.4	31.5	3.7	50.8	2.8	12
Forest	2009	3.8	5.8	13.9	15.2	24.3	23.3
Forest	2010	5	10	5.6	11.3	12.2	11.9
Forest	2011	1	26.2	4.8	29.9	6.4	18.7
Shrub/Scrub	2009	1.5	8.1	12.2	18.1	15.8	23.3
Shrub/Scrub	2010	4.3	11.8	5.9	4.5	17.9	11.1
Shrub/Scrub	2011	1.2	45.2	8.8	48.5	2.4	25.7
Grass/Pasture/Hay	2009	3.5	6.8	4.3	9.4	18.6	16
Grass/Pasture/Hay	2010	5	9.5	2	3.4	14.4	8
Grass/Pasture/Hay	2011	1.3	22.1	2.3	22.5	3.4	39.7
Cultivated crops	2009	4.3	10.7	6.1	11.5	12.8	16.8
Cultivated crops	2010	5.2	18.3	2.8	2.5	16.7	5.8
Cultivated crops	2011	3.1	37	3.1	21.4	1.8	17.3
Wetlands	2009	3	15.6	14.2	18.1	21	15.5
Wetlands	2010	4.7	15.2	5.7	12.8	14.5	11.8
Wetlands	2011	2.3	48.8	5	62.4	3.3	15

Table 8. Areas of land (square kilometers) in the land-cover groups and National Climate Assessment (NCA) regions during the growing season (original analysis).

Land cover group	Northeast	Southeast	Midwest	Great Plains	Northwest	Southwest	Total
Developed	29,424	54,489	43,434	25,347	7,896	20,046	180,636
Forest	381,616	576,589	352,536	213,895	248,853	281,702	2,055,192
Shrub/Scrub	4,122	22,780	692	425,991	207,433	743,434	1,404,451
Grass/Pasture/Hay	46,290	156,868	130,613	911,984	36,705	201,615	1,484,075
Cultivated crops	27,317	162,990	544,719	450,627	60,256	74,322	1,320,230
Wetlands	14,056	141,715	57,866	27,902	3,021	2,612	247,172
Total	502,825	1,115,431	1,129,860	2,055,746	564,164	1,323,731	6,691,756

Table 9. Condition of major land-cover types in the Northeast National Climate Assessment (NCA) region for 2009–11(original analysis).

Land Cover Class	Year	Drought	Normal	Moist
Developed	2009	6.66	84.08	9.26
Developed	2010	4.23	78.33	17.45
Developed	2011	2.37	59.62	38.01
Forest	2009	3.78	87.2	9.02
Forest	2010	5.04	87.43	7.53
Forest	2011	1.02	46.26	52.72
Shrub/Scrub	2009	1.49	85.16	13.36
Shrub/Scrub	2010	4.34	89.25	6.41
Shrub/Scrub	2011	1.23	34.11	64.67
Grass/Pasture/Hay	2009	3.47	90.33	6.2
Grass/Pasture/Hay	2010	5.02	88.6	6.37
Grass/Pasture/Hay	2011	1.27	48.39	50.33
Cultivated crops	2009	4.32	88.52	7.16
Cultivated crops	2010	5.2	83.17	11.64
Cultivated crops	2011	3.12	52.64	44.24
Wetlands	2009	2.99	81.07	15.94
Wetlands	2010	4.69	75.18	20.14
Wetlands	2011	2.27	49.73	48.01

Table 10. Condition of major land-cover types in the Southeast National Climate Assessment (NCA) region for 2009–11 (original analysis).

Land Cover Class	Year	Drought	Normal	Moist
Developed	2009	11.15	80.43	8.43
Developed	2010	8.83	75.19	15.98
Developed	2011	31.54	61.79	6.67
Forest	2009	5.75	83.64	10.61
Forest	2010	10.03	78.66	11.31
Forest	2011	26.15	65.53	8.32
Shrub/Scrub	2009	8.07	81.29	10.64
Shrub/Scrub	2010	11.76	74.73	13.52
Shrub/Scrub	2011	45.2	52.41	2.4
Grass/Pasture/Hay	2009	6.79	82.7	10.52
Grass/Pasture/Hay	2010	9.45	77.35	13.2
Grass/Pasture/Hay	2011	22.05	66.51	11.44
Cultivated crops	2009	10.72	78.25	11.03
Cultivated crops	2010	18.29	73.48	8.22
Cultivated crops	2011	37.02	55.35	7.63
Wetlands	2009	15.63	77.61	6.76
Wetlands	2010	15.22	74.2	10.57
Wetlands	2011	48.75	49.38	1.87

Table 11. Condition of major land-cover types in the Midwest National Climate Assessment (NCA) region for 2009–11 (original analysis).

Land Cover Class	Year	Drought	Normal	Moist
Developed	2009	6.75	72.56	20.69
Developed	2010	4.76	79.38	15.86
Developed	2011	3.7	60.25	36.05
Forest	2009	13.95	76.91	9.14
Forest	2010	5.62	80.81	13.58
Forest	2011	4.78	74.17	21.06
Shrub/Scrub	2009	12.18	82.12	5.7
Shrub/Scrub	2010	5.94	84.36	9.7
Shrub/Scrub	2011	8.82	80.43	10.75
Grass/Pasture/Hay	2009	4.33	76.5	19.18
Grass/Pasture/Hay	2010	2	61.27	36.73
Grass/Pasture/Hay	2011	2.29	78.42	19.29
Cultivated crops	2009	6.15	78.17	15.69
Cultivated crops	2010	2.77	67.59	29.64
Cultivated crops	2011	3.14	72.9	23.97
Wetlands	2009	14.21	74.73	11.07
Wetlands	2010	5.74	80.86	13.4
Wetlands	2011	4.99	77.96	17.05

Table 12. Condition of major land-cover types in the Great Plains National Climate Assessment (NCA) region for 2009–11 (original analysis).

Land Cover Class	Year	Drought	Normal	Moist
Developed	2009	17.92	75.12	6.96
Developed	2010	4.27	77.53	18.2
Developed	2011	50.75	45.39	3.87
Forest	2009	15.2	77.96	6.83
Forest	2010	11.27	80.01	8.72
Forest	2011	29.92	56.75	13.33
Shrub/Scrub	2009	18.14	75.18	6.69
Shrub/Scrub	2010	4.53	75.29	20.17
Shrub/Scrub	2011	48.55	38.17	13.28
Grass/Pasture/Hay	2009	9.39	79.71	10.9
Grass/Pasture/Hay	2010	3.43	74.46	22.11
Grass/Pasture/Hay	2011	22.52	53.17	24.31
Cultivated crops	2009	11.45	77.49	11.06
Cultivated crops	2010	2.53	69.89	27.58
Cultivated crops	2011	21.44	58.35	20.21
Wetlands	2009	18.1	75.36	6.54
Wetlands	2010	12.79	74.08	13.13
Wetlands	2011	62.39	29.83	7.79

Table 13. Condition of major land-cover types in the Northwest National Climate Assessment (NCA) region for 2009–11 (original analysis).

Land Cover Class	Year	Drought	Normal	Moist
Developed	2009	24.37	73.59	2.03
Developed	2010	6.43	85.76	7.81
Developed	2011	2.77	69.92	27.31
Forest	2009	24.25	72.48	3.27
Forest	2010	12.2	80.71	7.09
Forest	2011	6.38	73.79	19.84
Shrub/Scrub	2009	15.81	78.87	5.33
Shrub/Scrub	2010	17.92	78.5	3.57
Shrub/Scrub	2011	2.41	72.97	24.62
Grass/Pasture/Hay	2009	18.6	76.29	5.1
Grass/Pasture/Hay	2010	14.35	81.22	4.43
Grass/Pasture/Hay	2011	3.41	73.13	23.46
Cultivated crops	2009	12.83	79.17	8
Cultivated crops	2010	16.7	78.58	4.72
Cultivated crops	2011	1.84	70.56	27.59
Wetlands	2009	20.99	73.24	5.77
Wetlands	2010	14.45	81.03	4.52
Wetlands	2011	3.26	72.77	23.97

Table 14. Condition of major land-cover types in the Southwest National Climate Assessment (NCA) region for 2009–11 (original analysis).

Land Cover Class	Year	Drought	Normal	Moist
Developed	2009	22.68	74.06	3.26
Developed	2010	8.48	83.16	8.35
Developed	2011	12.02	68.83	19.15
Forest	2009	23.31	73.27	3.42
Forest	2010	11.94	80.51	7.55
Forest	2011	18.68	67.74	13.58
Shrub/Scrub	2009	23.27	74.31	2.42
Shrub/Scrub	2010	11.05	82.81	6.14
Shrub/Scrub	2011	25.65	63.7	10.65
Grass/Pasture/Hay	2009	16.01	77.22	6.77
Grass/Pasture/Hay	2010	7.98	78.5	13.52
Grass/Pasture/Hay	2011	39.7	53.77	6.53
Cultivated crops	2009	16.78	73	10.23
Cultivated crops	2010	5.81	78.22	15.97
Cultivated crops	2011	17.32	62.16	20.52
Wetlands	2009	15.54	80.25	4.21
Wetlands	2010	11.75	80.78	7.47
Wetlands	2011	14.99	68.12	16.89

The percentages of the land area in drought reflect different impacts of drought for the analyzed land-cover types in the NCA regions. When the NCA regions are compared, results indicate that the drought in 2011 was particularly strong in the shrub/scrub areas of the Great Plains (48.5 percent of 425,991 km^2) (fig. 6) and the cultivated crops of the Southeast (37.0 percent of 162,990 km^2) (fig. 7). The cultivated crops in 2011 in the Great Plains and Southwest regions also exhibited strong drought influences (21.4 percent and 17.3 percent, respectively) (table 7). The Grassland/pasture/hay category had major drought effects in 2011 in several regions (39.7 percent in the Southwest, 22.5 percent in the Great Plains, and 22.1 percent in the Southeast). Drought affected 48.8 percent of the 141,715 km^2 of wetlands in the Southeast. Some categories experienced high percentages of land with drought influence, even though the total area of land was relatively small (for example, 62.4 percent of 27,902 km^2 of wetlands in the Great Plains). Summarizing drought effects across large regions will moderate the drought extremes that have occurred in smaller regions. For example, in 2011, the northern Great Plains were exceptionally wet while the southern tier experienced a record-breaking drought.

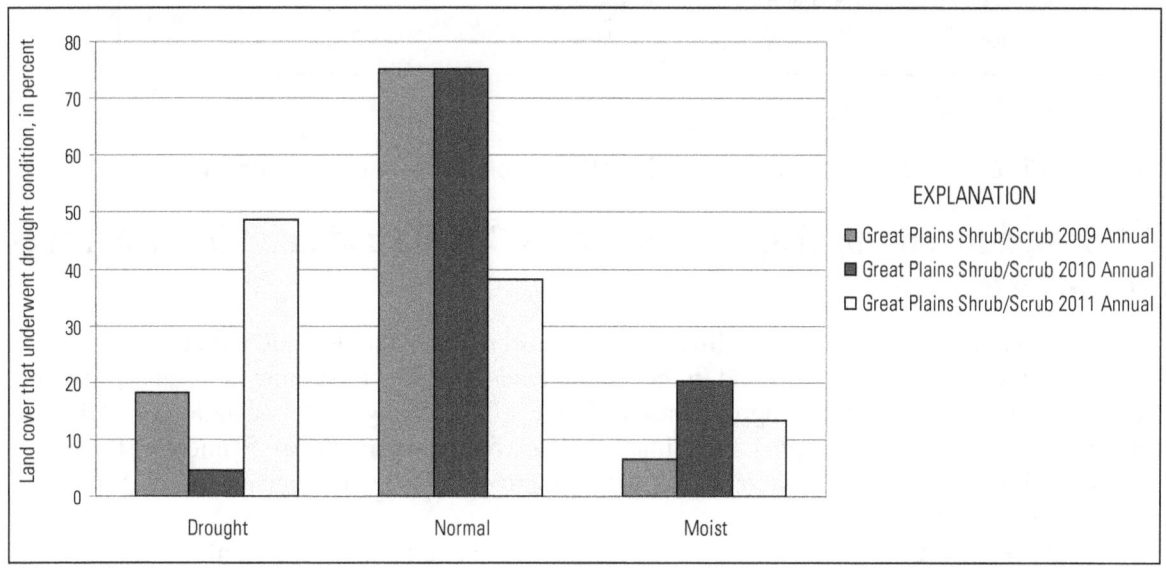

Figure 6.   Shrub/scrub drought trends in the Great Plains region from 2009–11 (original analysis).

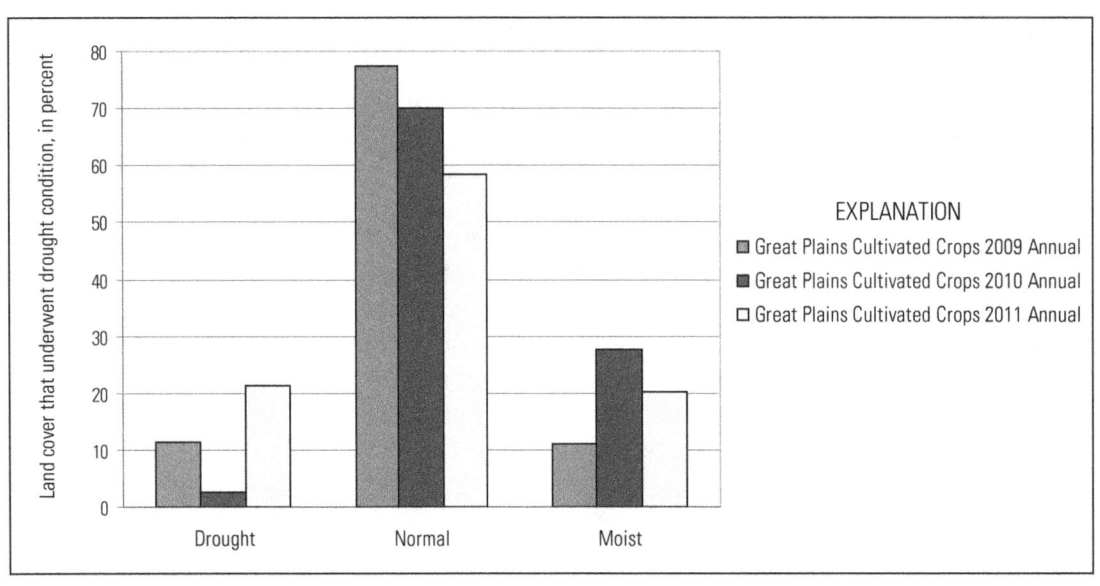

Figure 7. Cultivated crops drought trends in the Great Plains from 2009–11 (original analysis).

5.0 A Design for an Ongoing Assessment of Climate and Land-Use and Land-Cover Impacts

The goal of an assessment capability should focus on understanding and explaining how climate and LULC influence each other now and in the future. Understanding the connections and consequences of LULC and climate will require ongoing monitoring of LULC change, the translation of LULC into parameters relevant to meteorological and climatological processes, and the assessment of the real impacts that climate and LULC have on each other. An ongoing LULC climate assessment should address the following:

- What are the primary contemporary trends in land use and land cover that affect, or are affected by, weather and climate?
- Of these trends, which sectors and regions are most affected by weather and climate variability?
- What land uses and regions are most vulnerable to climate change?
- How are land-use practices adapting to climate change?

Assessment capabilities should include the means to evaluate the interactions of land use and management with climate change in a way that will help decision makers mitigate or adapt to the changes. This would require general principles on how the management may need to adapt in the context of changing climate rather than working from implicit assumptions on static climatic conditions (West and others, 2009). In particular, land managers and others may need to assess whether changes in climate would push advantageous LULC conditions beyond the point that they provide the necessary goods and services. Future LULC management strategies may require methods that incorporate an evaluation of how climate and LULC change effects can combine to influence a wide range of social and economic benefits as well as ecological factors such as the migration of species, shifting mosaics of wetlands, and disturbances on climate-biology relations. The mechanisms of ecological response need to be incorporated into the design of the monitoring systems (Beever and Woodward, 2011b).

5.1 Assessment Research and Development Needs

In order to address these questions, research and development may need to focus on a series of objectives that ensure assessment results are credible and relevant. Some specific foundational objectives for an ongoing assessment capability include the following:

1. Improve the understanding of the connections between LULC and weather and climate. For effective modeling and assessment of climate and LULC forcings and feedbacks, research should be carried out that:

    a. Improves the current understanding of how LULC and atmospheric interactions are linked at local to global scales.

    b. Validates these connections through an analysis of the historical record. Past changes could be identified in LULC that are attributable to changes in climate in order to project future changes in LULC that could result from changes in climate.

Both the climate system and the human activities that result in LULC changes are complex systems that can only be observed with limited direct observations. For understanding both systems and how they interact, it is necessary to use models. Currently, many of the modeling systems are run to simulate the climate system with assumptions on how land-cover patterns will change with time (Brovkin et al., 2006) or model land-cover change with assumptions on how climate will change with time (Sohl and Sayler, 2008). A current research objective is to better couple these models so that feedback between the systems can be incorporated.

Any given model may be most appropriate for a limited range in space and time; so it may be necessary to use a suite of models to fully test system behaviors. The models are calibrated using observational data in historical periods, and the calibrated models may be used to project into the future (with increasing uncertainty through time). The ultimate purpose for the data collection and model simulations is to inform decision makers and the public about anticipated impacts on human and ecological systems (including agricultural, forest, wildlife, and human communities), so that activities to mitigate and adapt to the changes can be planned and undertaken.

In order to provide input to climate and meteorological models, LULC forecast models must be spatially explicit, provide a means to parameterize key land-atmosphere interactions, and provide scenario-based forecasts for 50–100 years. There are several forecast models in use that can be used to project regional to national LULC patterns into the future (for example, FORE–SCE by Sohl and others, 2010 that is being used for the USGS LandCarbon study on carbon management opportunities), but none of the models provide more than rudimentary handling of climate-model parameters. The most viable future LULC projections will be those based on well-defined and vetted scenarios. Current Federal agency scenario research has focused on using the IPCC SRES storylines to set the stage for downscaled land-change forecasts in the United States. The EPA Global Change Research Program's ICLUS project has established scenarios broadly consistent with the global-scale IPCC SRES storylines of population growth and economic development (U.S. Environmental Protection Agency, 2009). The EPA ICLUS forecast provides spatially explicit maps of housing density and the expansion of impervious surfaces based on SRES storylines.

Because the SRES storylines were developed for use by climate change modelers to develop projections of future climate, they represent a reasonable starting point for a LULC climate assessment in the United States. The USGS has also used SRES storylines to establish a nationally consistent library of future land change scenarios for use in addressing biological carbon sequestration opportunities (Zhu and others, 2010). The USGS effort addresses all major land-cover types found in the conterminous United States and is being developed for the ecoregions of the country. The SRES storylines provide the broad-level boundary conditions in the United States, and historical land-cover trends are used to establish the basis for future regional land changes.

Improvements in land-change forecasts should benefit from the recently commissioned National Research Council (NRC) study on the Needs and Research for Land-Change Modeling, which is an important step toward improving future LULC forecasts. This study was recommended by the USGCRP Land Use Interagency Working group based on the LULC change science priorities specified in the 2003 CCSP science strategy. The NRC study should provide a thorough review of the present status of spatially explicit land-change modeling approaches and describe future data and research needs so that model outputs can better assist the science, policy, and decision-support communities.

2. Improve the coupling of LULC states and conditions within meteorological and climatological models. This will require:

   a. The translation of LULC variables into quantitative parameters that directly relate to the physics and chemistry connected with the exchange of energy, water, and momentum between the land surface and the atmosphere.

   b. The ongoing development of multi-resolution LULC parameters needed to improve the accuracy of simulation models forecasts. LULC model parameters should become more current and accessible. This involves the ongoing development of Climate Data Records and Essential Climate Variables—data sets based on international standards that ensure relevant, stable measures needed to understand climate and climate impacts are developed (Global Climate Observing system, 2010).

   c. Efforts to couple climate and LULC forecast models so that the dynamics of each component are part of the modeling process.

   d. Model inter-comparability studies should be used to determine strengths and weakness of different models and modeling approaches.

Over the years, significant progress has been made in model design (physics and chemistry) that can address LULC changes and their interactions with weather and climate. Subsequently, both regional and global-scale modeling efforts have been undertaken to determine the impacts and interactions. However, it is evident that experimental design and modeling capabilities need further improvement.

A number of in-situ and remote weather and climate observation platforms are currently available that can be used to identify signals of impacts of LULC change on atmospheric data. These include: U.S. Climate Reference Network (USCRN), new U.S. Regional Climate Reference Network (USRCRN), and regional mesonets. The latter two could be excellent platforms for regional and local-scale signals. In addition, satellite data can be used with the in-situ observations when available.

The International Global Energy and Water Cycle Experiment (GEWEX) is an example of a large collaborative research campaign in which many scientific organizations interact to achieve broad and integrative scientific goals. One such goal is to understand how land surface hydrology influences water availability and security (Trenbreth, 2011). One element of reaching the goal is to account for realistic land-surface complexity, including human influences such as land-use change and urbanization. Water quality, including water temperature and nutrient loadings, is affected by industrial and power plant use (National Research Council, 2001). The availability of water for human use and ecological systems will be affected by ecosystem responses to projected changing climate. Extremes of weather can cause water systems to be vulnerable, and good management and governance can increase resilience.

A range of multiscale geospatial land-cover-related datasets would cover the range of analysis functions—for example, model parameterization, land-condition monitoring, and impact evaluation.

Earth observations from global daily polar orbiter instruments such as the NOAA Advanced Very High Resolution Radiometers (AVHRR), NASA Moderate Resolution Imaging Spectroradiometer (MODIS) and NASA–NOAA Visible Infrared Imager Radiometer Suite (VIIRS) can provide current land parameters used in meteorological models (for example, surface albedo, surface temperature, leaf area index, and land cover). The current MODIS land products provide an important source for parameter datasets (Justice and others, 2002). These same data can also be used to monitor weekly changes in vegetation condition. The VegDRI product used earlier in this report to evaluate 2009–11 drought impacts on land cover is an example of a global daily imaging land condition product (Brown and others, 2008). Validation must be a component of all data activities.

Higher resolution land-cover characteristics and land-cover-change data are needed to identify and evaluate the specific cover types that are affected by weather and climate variability. Landsat-scale (30 m resolution) such as the NLCD dataset is suited to this application. However, an operational assessment would require land-cover information at a more frequent interval than NLCD. The planned USFS-USGS Land Cover Monitoring System concept that would provide Landsat-scale annual change data is a stronger, potential long-term candidate. Finally, land-use data are needed to understand impacts and mitigation opportunities. Sectoral products, such as Forest Inventory and Analysis data, are useful, but the absence of spatially explicit national land-use data is an issue.

New data sources and analysis techniques that have been applied to land-cover analysis possibly could be extended to incorporate time series of both land surface and climate observations and then could be used to analyze the interactions of LULC and climate (Knorn and others, 2009; Huang and others, 2010; Kennedy and others, 2010; Roy and others, 2010). Knowledge of such interactions could be used to assess the feedbacks between the land cover and climate systems and incorporated into the next generation of coupled LULC and climate models.

3. Increase the understanding of the relations between climate and LULC impacts. This would require understanding how:

   a. Weather and climate variables differentially affect the different LULC types.

   b. Different landscape variables (for example, ecoregions, topography, and land ownership) modify these relations.

   c. LULC and ecosystems recover after disturbance events.

Specialties under various academic disciplines study LULC changes and their interactions with the weather and climate. Many of these groups have started to increase scholarly communications among themselves, which has enhanced the flow of knowledge in the recent years. However, further concerted efforts to expand these collaborations would provide maximum benefit to society.

4. An assessment capability is needed that provides ongoing information on the impacts of weather conditions and climate trends on LULC. Reporting should focus on providing a clear understanding of the economic, social, and ecological impacts, and on the ways LULC changes in response to events and trends. Specific considerations include:

   a. Distinguishing between short- and long-term climate patterns and their impacts on LULC. This will allow decision makers at all levels to determine mitigation or coping mechanisms.

   b. Establishing the capability to evaluate the impacts of extreme weather events on LULC, addressing the stresses on economic, social, and ecological systems.

   c. Providing an explanation on how local and regional climate and LULC impacts affect national and global economic, social, and ecological systems.

d. Providing forecasts on the potential land-use impacts because of weather events and climate trends. These should include information on mitigation options and coping mechanisms.

e. Describing how LULC changes related to weather and climate affect social and economic systems (for example, impacts on forest productivity, shortened/lengthened growing seasons for crop production). This should include the impacts of climate-induced LULC change on people's livelihoods.

Regions that are currently experiencing rapid LULC changes and other ecologically sensitive and vulnerable areas could further be considered for weather and climate monitoring to better understand the pathways, mechanisms, and processes related to LULC change impacts on the atmosphere. Specifically, in addition to existing observation platforms, establishing weather and climate monitoring capabilities needs to be considered in some of the above noted areas. This effort could be completed in phases.

The combination of LULC and climate may contribute to assessment of impacts of the capability of LULC systems to provide future goods and services. In order to detect evidence of land degradation, reference conditions can be used to determine deviation from a sustainable state. For example, Herrick and others (2010) show how data from the National Resources Inventory, along with remotely sensed imagery, soil surveys, and climate models, can be used to stratify landscapes in a way that allows the definition of reference conditions based on the long-term ecological potential of the land. Deviation from the reference conditions indicated possible land degradation.

Although it is difficult to quantify the societal benefits of LULC practices and conditions, some research is providing a basis to understand these benefits by using more traditional measures of economic output. For example, Nelson and others (2009) have developed a modeling tool to predict changes in ecosystem services, biodiversity conservation, and commodity production levels and have applied the tool to the Willamette Basin in Oregon.

5. Maintaining an outreach capability that ensures rapid access to assessment inputs and outputs, interpretation of the results, and technical support for decision makers and scientists engaged in climate and LULC issues.

The communication of results and applications services used in assessing LULC and climate is critical to ensure the effective use of data, models, and analyses. This could include the provision of decision support tools tailored to the particular needs of different stakeholder groups.

5.2 Institutional Capabiltiies

An ongoing assessment of LULC and weather and climate connections would require the participation of a number of USGCRP members as well as academic and industry researchers. Key agency participants and contributions could include the following:

EPA – land change scenario information,

NASA – land surface parameterizations derived from remote sensing, land cover and atmospheric research ,

NOAA – meteorological and climatological expertise, atmospheric observations, in situ instrument records,

NSF – research support, participation of key observation networks such as NEON,

USDA – a wide range of datasets and assessments, including USFS Forest Inventory and Analysis and forest cover products, Natural Resources Conservation Service National Resources Inventory data, National Agricultural Statistical Service annual crop area and type data and agricultural census results, and

USGS – land-cover map products, land-change scenarios, land-change modeling and forecasts, climate and land-use research.

The integration of Federal capabilities will be challenging and will require a fresh approach.

Beever and Woodward (2011a), for example, suggest that climate and land monitoring that effectively supports resource management might ideally be structured to address the actual spatial and temporal scales of relevant processes, rather than the artificial boundaries of individual land-management units.

In summary, assessment and monitoring systems for understanding the changing relations between land use, land cover, and climate should make use of information from multiple scales of space and time. Field plot measurements, such as the FIA and NRI, flux towers for detailed understanding of carbon dynamics, soil surveys, and long-term ecological research sites are needed for detailed understanding of processes. Remotely sensed data, particularly the Landsat time series integrated with higher and lower resolution data, can provide information on the spatial magnitude and directions of change. Integration of these data sources in models allows bridging the gaps in observation across space and time, and allow simulation of processes that are not directly observable. Testing of multiple scenarios may make it possible to separate the influences of different processes (for example, land management compared to climate change or weather extremes) as they influence ecosystems and human activities. New methods of analysis, in which entire time series of images can be analyzed at once, provide new possibilities in classifying land-cover change. It is possible that the algorithms used in such analyses could be adapted to analyze joint time series of climate change, weather extremes, and land-cover change to separate and investigate the interactions of these variables. Knowledge of such interactions could be coded into coupled models of the climate and LULC interactions could be used to forecast scenarios of future system behavior. Such forecasts could help identify critical weaknesses in existing planning for mitigation and adaptation. The assessment system should include continued contact with groups that represent decision makers for urban and regional planning, agricultural and forest land management, biodiversity conservation, and ecological research, so that the models are sensitive to the types of policy choices that will be needed in the future. The research should be coordinated with national and international campaigns that have complementary interests, such as National Ecological Observatory Network (NEON), Global Energy and Water Balance Experiment (GEWEX), and the Global Earth Observation System of Systems (GEOSS).

The basic inputs needed for an operational LULC-climate assessment capability are available, but some inputs will require some improvements or changes in specifications in order to provide the timeliness and geographic coverage required. The real challenge will be the identification of a Federal host to lead the assessment process. NOAA, USDA, or the USGS are the logical candidates based on their mission objectives and current investments in climate and LULC.

6.0 Concluding Observations

Projected climate variability and change will challenge natural resources managers. Management challenges can be compromised when LULC change reduces, modifies, or complicates strategies for managing climate change. As a result, maintaining the societal and ecological benefits drawn from the Nation's land resources—land cover and land use— requires an integrated understanding of the bi-directional links between LULC and weather and climate. That understanding will likely come from an assessment capability assembled from the various USGCRP agency activities related to LULC and climate change. In this report, identification of the foundations of such a capacity—the state of science and the availability of the basic elements that might be included in a monitoring and assessment activity—was begun.

The foundation of climate-LULC understanding, the scientific investigations addressing those bidirectional links, is growing rapidly. Studies on the basic mechanics and processes governing the exchange of water and energy between the land surface and the atmosphere are maturing and being used in both experimental and operational forecasting. Years of land-atmosphere interaction research has led to significant maturity in the ability to analyze the role land cover plays in weather and climate formation. The contemporary land-cover parameters used in those models are also available. As a result, it is increasingly feasible to simulate local to regional LULC influences on weather and climate formation. The simulation of future LULC-climate connections is more complicated because most future LULC projections are not dynamically linked to climate and meteorology models. This is an area where more research and development are needed.

The national investments in Earth observations can provide the means to identify LULC stresses, to map the condition of LULC across the Nation, and determine how different regions are changing or adapting to different weather and climate conditions. Where there is considerable capacity to provide near real-time monitoring of LULC responses to climate, there is currently no systematic effort to monitor and evaluate those climate-driven LULC changes.

Considering the questions outlined in Section 5.0 that could drive a LULC-climate assessment capability, a general conclusion is that the basic elements needed for monitoring and assessment exist, though not necessarily in a format that is optimized for the assessment of LULC-climate impacts and feedbacks. The challenge will be the integration of capabilities, the enhancement of the different elements, and the maturing of assessment frameworks. Attention to the spatial and temporal scales of analysis, the geospatial framework for monitoring, assessing, and reporting LULC-climate issues, the detailed specifications and the validation of all inputs, outputs, and model assumptions would be needed. There are some obvious areas where improvements would be required. Most activities do not specifically address the issues occurring in Alaska, Hawaii, and the Territories of the United States. Most observational capabilities are more adept in monitoring croplands and forests rather than cities and rangelands. Efforts to integrate in situ and remotely sensed data would require more attention. Improvements in model coupling would be required. Perhaps most important for LULC assessments is the improvement in geospatial representation of land-use practices. This is problematic because land-use inputs are needed to address local social and economic issues. The next steps are to move beyond independent case studies and a rich assortment of technical tools and data into a designed, integrated framework for ongoing national assessments.

7.0 References

Adegoke, J.O., Pielke Sr., R.A., Eastman, J., Mahmood, R., and Hubbard, K.G., 2003, Impact of irrigation on midsummer surface fluxes and temperature under dry synoptic conditions—a regional atmospheric model study of the U.S. high plains: Monthly Weather Review, v. 131, no. 3, p. 556-564. (Also available online at *http://dx.doi.org/10.1175/1520-0493(2003)131<0556:ioioms>2.0.co;2.*)

Allard, J., and Carleton, A., 2010, Mesoscale associations between midwest land surface properties and convective cloud development in the warm season: Physical Geography, v. 31, no. 2, p. 107-136. (Also available online *at http://dx.doi.org/10.2747/0272-3646.31.2.107.*)

Anthes, R.A., 1984, Enhancement of convective precipitation by mesoscale variations in vegetative covering in semiarid regions: Journal of Climate and Applied Meteorology, v. 23, no. 4, p. 541-554. (Also available online at *http://dx.doi.org/10.1175/1520-0450(1984)023<0541:eocpbm>2.0.co;2.*)

Arnfield, A.J., 2003, Two decades of urban climate research—a review of turbulence, exchanges of energy and water, and the urban heat island: International Journal of Climatology, v. 23, no. 1, p. 1-26. (Also available online at *http://dx.doi.org/10.1002/joc.859.*)

Banner, E.B.K., Stahl, A.J., and Dodds, W.K., 2009, Stream discharge and riparian land use influence in-stream concentrations and loads of phosphorus from central plains watersheds: Environmental Management, v. 44, no. 3, p. 552-565. (Also available online at *http://dx.doi.org/10.1007/s00267-009-9332-6.*)

Barnston, A.G., and Schickedanz, P.T., 1984, The effect of irrigation on warm season precipitation in the southern Great Plains: Journal of Climate and Applied Meteorology, v. 23, no. 6, p. 865-888. (Also available online at *http://dx.doi.org/10.1175/1520-0450(1984)023<0865:teoiow>2.0.co;2.*)

Barrett, T.M., and Gray, A.N., 2011, Potential of a national monitoring program for forests to assess change in high-latitude ecosystems: Biological Conservation, v. 144, no. 5, p. 1285-1294. (Also available online at *http://dx.doi.org/10.1016/j.biocon.2010.10.015.*)

Beever, E.A., and Woodward, A., 2011, Design of ecoregional monitoring in conservation areas of high-latitude ecosystems under contemporary climate change: Biological Conservation, v. 144, no. 5, p. 1258-1269. (Also available online at *http://dx.doi.org/10.1016/j.biocon.2010.06.022.*)

Beever, E.A., and Woodward, A., 2011, Ecoregional-scale monitoring within conservation areas, in a rapidly changing climate: Biological Conservation, v. 144, no. 5, p. 1255-1257. (Also available online at *http://dx.doi.org/10.1016/j.biocon.2011.04.001.*)

Beltrán-Przekurat, A., Pielke Sr., R.A., Eastman, J.L., and Coughenour, M.B., 2011, Modelling the effects of land-use/land-cover changes on the near-surface atmosphere in southern South America: International Journal of Climatology. (Also available online at *http://dx.doi.org/10.1002/joc.2346.*)

Biggs, T.W., Scott, C.A., Gaur, A., Venot, J.-P., Chase, T., and Lee, E., 2008, Impacts of irrigation and anthropogenic aerosols on the water balance, heat fluxes, and surface temperature in a river basin: Water Resources Research, v. 44, no. 12, p. W12415. (Also available online at *http://dx.doi.org/10.1029/2008wr006847.*)

Blake, D., Brockett, P., Cox, S., and MacMinn, R., 2011, Longevity risk and capital markets—the 2009-2010 update: North American Actuarial Journal, v. 15, no. 2, p. 141-149. (Also available online at *http://www.soa.org/library/journals/north-american-actuarial-journal/2011/no-2/naaj-2011-vol15-no2.aspx.*)

Blate, G.M., Joyce, L.A., Littell, J.S., McNulty, S.G., Millar, C.I., Moser, S.C., Neilson, R.P., O'Halloran, K., and Peterson, D.L., 2009, Adapting to climate change in United States national forests: Unasylva, v. 60, no. 231-232, p. 57-62. (Also available online at *http://www.fao.org/docrep/011/i0670e/i0670e00.htm.*)

Bonan, G.B., 1997, Effects of land use on the climate of the United States: Climatic Change, v. 37, no. 3, p. 449-486. (Also available online at *http://dx.doi.org/10.1023/a:1005305708775.*)

Bonan, G.B., 2001, Observational evidence for reduction of daily maximum temperature by croplands in the midwest United States: Journal of Climate, v. 14, no. 11, p. 2430-2442. (Also available online at *http://dx.doi.org/10.1175/1520-0442(2001)014<2430:oefrod>2.0.co;2.*)

Bonfils, C., and Lobell, D., 2007, Empirical evidence for a recent slowdown in irrigation-induced cooling: Proceedings of the National Academy of Sciences, v. 104, no. 34, p. 13582-13587. (Also available online at *http://dx.doi.org/10.1073/pnas.0700144104.*)

Bradley, B.A., 2010, Assessing ecosystem threats from global and regional change—hierarchical modeling of risk to sagebrush ecosystems from climate change, land use and invasive species in Nevada, USA: Ecography, v. 33, no. 1, p. 198-208. (Also available online at *http://dx.doi.org/10.1111/j.1600-0587.2009.05684.x.*)

Broussard, W., and Turner, R.E., 2009, A century of changing land-use and water-quality relationships in the continental US: Frontiers in Ecology and the Environment, v. 7, no. 6, p. 302-307. (Also available online at *http://dx.doi.org/10.1890/080085.*)

Brovkin, V., Claussen, M., Driesschaert, E., Fichefet, T., Kicklighter, D., Loutre, M., Matthews, H., Ramankutty, N., Schaeffer, M., and Sokolov, A., 2006, Biogeophysical effects of historical land cover changes simulated by six Earth system models of intermediate complexity: Climate Dynamics, v. 26, no. 6, p. 587-600. (Also available online at *http://dx.doi.org/10.1007/s00382-005-0092-6.*)

Brovkin, V., Sitch, S., Von Bloh, W., Claussen, M., Bauer, E., and Cramer, W., 2004, Role of land cover changes for atmospheric CO_2 increase and climate change during the last 150 years: Global Change Biology, v. 10, no. 8, p. 1253-1266. (Also available online at *http://dx.doi.org/10.1111/j.1365-2486.2004.00812.x.*)

Brown, J., Wardlow, B., Tadesse, T., Hayes, M., and Reed, B., 2008, The Vegetation Drought Response Index (VegDRI)—a new integrated approach for monitoring drought stress in vegetation: GIScience & Remote Sensing, v. 45, no. 1, p. 16-46. (Also available online at *http://dx.doi.org/10.2747/1548-1603.45.1.16.*)

Burow, K.R., Nolan, B.T., Rupert, M.G., and Dubrovsky, N.M., 2010, Nitrate in groundwater of the United States, 1991-2003: Environmental Science and Technology, v. 44, no. 13, p. 4988-4997. (Also available online at *http://dx.doi.org/10.1021/es100546y.*)

Campra, P., Garcia, M., Canton, Y., and Palacios-Orueta, A., 2008, Surface temperature cooling trends and negative radiative forcing due to land use change toward greenhouse farming in southeastern Spain: Journal of Geophysical Research-Atmospheres, v. 113, no. D18, p. D18109. (Also available online at *http://dx.doi.org/10.1029/2008jd009912.*)

Canadell, J.G., Le Quéré, C., Raupach, M.R., Field, C.B., Buitenhuis, E.T., Ciais, P., Conway, T.J., Gillett, N.P., Houghton, R.A., and Marland, G., 2007, Contributions to accelerating atmospheric CO_2 growth from economic activity, carbon intensity, and efficiency of natural sinks: Proceedings of the National Academy of Sciences, v. 104, no. 47, p. 18866-18870. (Also available online at *http://dx.doi.org/10.1073/pnas.0702737104.*)

Carleton, A.M., Arnold, D.L., Travis, D.J., Curran, S., and Adegoke, J.O., 2008, Synoptic circulation and land surface influences on convection in the midwest U.S. "corn belt" during the summers of 1999 and 2000. Part I—composite synoptic environments: Journal of Climate, v. 21, no. 14, p. 3389-3415. (Also available online at *http://dx.doi.org/10.1175/2007jcli1578.1.*)

Carleton, A.M., Travis, D.J., Adegoke, J.O., Arnold, D.L., and Curran, S., 2008, Synoptic circulation and land surface influences on convection in the midwest U.S. "corn belt" during the summers of 1999 and 2000. Part II—role of vegetation boundaries: Journal of Climate, v. 21, no. 15, p. 3617-3641. (Also available online at *http://dx.doi.org/10.1175/2007jcli1584.1.*)

Chen, J., Avise, J., Guenther, A., Wiedinmyer, C., Salathe, E., Jackson, R.B., and Lamb, B., 2009, Future land use and land cover influences on regional biogenic emissions and air quality in the United States: Atmospheric Environment, v. 43, no. 36, p. 5771-5780. (Also available online at *http://dx.doi.org/10.1016/j.atmosenv.2009.08.015.*)

Choi, W., 2008, Catchment-scale hydrological response to climate-land-use combined scenarios—a case study for the Kishwaukee River basin, Illinois: Physical Geography, v. 29, no. 1, p. 79-99. (Also available online at *http://dx.doi.org/10.2747/0272-3646.29.1.79.*)

Christy, J.R., Norris, W.B., Redmond, K., and Gallo, K.P., 2006, Methodology and results of calculating central California surface temperature trends—evidence of human-induced climate change?: Journal of Climate, v. 19, no. 4, p. 548-563. (Also available online at *http://dx.doi.org/10.1175/jcli3627.1.*)

Chuang, T.W., Hildreth, M.B., Vanroekel, D.L., and Wimberly, M.C., 2011, Weather and land cover influences on mosquito populations in Sioux Falls, South Dakota: Journal of Medical Entomology, v. 48, no. 3, p. 669-679. (Also available online at *http://dx.doi.org/10.1603/ME10246.*)

Claussen, M., Brovkin, V., and Ganopolski, A., 2001, Biogeophysical versus biogeochemical feedbacks of large-scale land cover change: Geophysical Research Letters, v. 28, no. 6, p. 1011-1014. (Also available online at *http://dx.doi.org/10.1029/2000gl012471.*)

Climate Change Science Program, 2003, Strategic plan for the U.S. climate change science program: Washington, D.C., 202 p. (Also available online at *http://purl.access.gpo.gov/GPO/LPS64573.*)

Costa, M.H., Yanagi, S.N.M., Souza, P.J.O.P., Ribeiro, A., and Rocha, E.J.P., 2007, Climate change in Amazonia caused by soybean cropland expansion, as compared to caused by pastureland expansion: Geophysical Research Letters, v. 34, no. 7, p. L07706. (Also available online at *http://dx.doi.org/10.1029/2007gl029271.*)

Cuo, L., Lettenmaier, D.P., Alberti, M., and Richey, J.E., 2009, Effects of a century of land cover and climate change on the hydrology of the Puget Sound basin: Hydrological Processes, v. 23, no. 6, p. 907-933. (Also available online at *http://dx.doi.org/10.1002/hyp.7228.*)

Dale, V.H., 1997, The relationship between land-use change and climate change: Ecological Applications, v. 7, no. 3, p. 753-769. (Also available online at *http://dx.doi.org/10.1890/1051-0761(1997)007[0753:trbluc]2.0.co;2.*)

Dale, V.H., Efroymson, R.A., and Kline, K.L., 2011, The land use-climate change-energy nexus: Landscape Ecology, v. 26, no. 6, p. 755-773. (Also available online at *http://dx.doi.org/10.1007/s10980-011-9606-2.*)

Davin, E.L., and de Noblet-Ducoudré, N., 2010, Climatic impact of global-scale deforestation—radiative versus nonradiative processes: Journal of Climate, v. 23, no. 1, p. 97-112. (Also available online at *http://dx.doi.org/10.1175/2009jcli3102.1.*)

Davin, E.L., de Noblet-Ducoudré, N., and Friedlingstein, P., 2007, Impact of land cover change on surface climate—relevance of the radiative forcing concept: Geophysical Research Letters, v. 34, no. 13, p. L13702. (Also available online at *http://dx.doi.org/10.1029/2007gl029678.*)

DeAngelis, D., 2010, Foreword to the Siberian lakes special issue: Aquatic Ecology, v. 44, no. 3, p. 479-479. (Also available online at *http://dx.doi.org/10.1007/s10452-010-9337-5.*)

Defries, R.S., Bounoua, L., and Collatz, G.J., 2002, Human modification of the landscape and surface climate in the next fifty years: Global Change Biology, v. 8, no. 5, p. 438-458. (Also available online at *http://dx.doi.org/10.1046/j.1365-2486.2002.00483.x.*)

Delgado, J.A., Groffman, P.M., Nearing, M.A., Goddard, T., Reicosky, D., Lal, R., Kitchen, N.R., Rice, C.W., Towery, D., and Salon, P., 2011, Conservation practices to mitigate and adapt to climate change: Journal of Soil and Water Conservation, v. 66, no. 4, p. 118A-129A. (Also available online at *http://dx.doi.org/10.2489/jswc.66.4.118A.*)

DellaSala, D.A., Karr, J.R., and Olson, D.M., 2011, Roadless areas and clean water: Journal of Soil and Water Conservation, v. 66, no. 3, p. 78A-84A. (Also available online at *http://dx.doi.org/10.2489/jswc.66.3.78A.*)

Douglas, E.M., Beltrán-Przekurat, A., Niyogi, D., Pielke Sr., R.A., and Vörösmarty, C.J., 2009, The impact of agricultural intensification and irrigation on land–atmosphere interactions and Indian monsoon precipitation—a mesoscale modeling perspective: Global and Planetary Change, v. 67, no. 1–2, p. 117-128. (Also available online at *http://dx.doi.org/10.1016/j.gloplacha.2008.12.007.*)

Douglas, E.M., Niyogi, D., Frolking, S., Yeluripati, J.B., Pielke Sr., R.A., Niyogi, N., Vörösmarty, C.J., and Mohanty, U.C., 2006, Changes in moisture and energy fluxes due to agricultural land use and irrigation in the Indian monsoon belt: Geophysical Research Letters, v. 33, no. 14, p. L14403. (Also available online at *http://dx.doi.org/10.1029/2006gl026550.*)

Drummond, M.A., Auch, R.F., Karstensen, K.A., Sayler, K.L., Taylor, J.L., and Loveland, T.R., 2012, Land change variability and human–environment dynamics in the United States Great Plains: Land Use Policy, v. 29, no. 3, p. 710-723. (Also available online at *http://dx.doi.org/10.1016/j.landusepol.2011.11.007.*)

Drummond, M.A., and Loveland, T.R., 2010, Land-use pressure and a transition to forest-cover loss in the eastern United States: BioScience, v. 60, no. 4, p. 286-298. (Also available online at *http://dx.doi.org/10.1525/bio.2010.60.4.7.*)

Duniway, M.C., Herrick, J.E., Pyke, D.A., and Toledo P, D., 2010, Assessing transportation infrastructure impacts on rangelands—test of a standard rangeland assessment protocol: Rangeland Ecology and Management, v. 63, no. 5, p. 524-536. (Also available online at *http://dx.doi.org/10.2111/REM-D-09-00176.1.*)

Easterling, W.E., 1996, Adapting North American agriculture to climate change in review: Agricultural and Forest Meteorology, v. 80, no. 1, p. 1-53. (Also available online at *http://dx.doi.org/10.1016/0168-1923(95)02315-1.*)

Eglin, T., Ciais, P., Piao, S.L., Barre, P., Bellassen, V., Cadule, P., Chenu, C., Gasser, T., Koven, C., Reichstein, M., and Smith, P., 2010, Historical and future perspectives of global soil carbon response to climate and land-use changes: Tellus, Series B: Chemical and Physical Meteorology, v. 62, no. 5, p. 700-718. (Also available online at *http://dx.doi.org/10.1111/j.1600-0889.2010.00499.x.*)

Fall, S., Diffenbaugh, N.S., Niyogi, D., Pielke Sr., R.A., and Rochon, G., 2010, Temperature and equivalent temperature over the United States (1979-2005): International Journal of Climatology, v. 30, no. 13, p. 2045-2054. (Also available online at *http://dx.doi.org/10.1002/joc.2094.*)

Feddema, J.J., Oleson, K.W., Bonan, G.B., Mearns, L.O., Buja, L.E., Meehl, G.A., and Washington, W.M., 2005, The importance of land-cover change in simulating future climates: Science, v. 310, no. 5754, p. 1674-1678. (Also available online at *http://dx.doi.org/10.1126/science.1118160.*)

Ford, C.R., Laseter, S.H., Swank, W.T., and Vose, J.M., 2011, Can forest management be used to sustain water-based ecosystem services in the face of climate change?: Ecological Applications, v. 21, no. 6, p. 2049-2067. (Also available online at *http://dx.doi.org/10.1890/10-2246.1.*)

Fry, J.A., Xian, G., Jin, S., Dewitz, J.A., Homer, C.G., Yang, L., Barnes, C.A., Herold, N.D., and Wickham, J.D., 2011, Completion of the 2006 National Land Cover Database for the conterminous United States: Photogrammetric Engineering and Remote Sensing, v. 77, no. 9, p. 858-864. (Also available online at *http://asprs.org/Photogrammetric-Engineering-and-Remote-Sensing/PE-RS-Journals.html.*)

Gallant, A.L., Sadinski, W., Roth, M.F., and Rewa, C.A., 2011, Changes in historical Iowa land cover as context for assessing the environmental benefits of current and future conservation efforts on agricultural lands: Journal of Soil and Water Conservation, v. 66, no. 3, p. 67A-77A. (Also available online at *http://dx.doi.org/10.2489/jswc.66.3.67A.*)

Gameda, S., Qian, B., Campbell, C.A., and Desjardins, R.L., 2007, Climatic trends associated with summerfallow in the Canadian prairies: Agricultural and Forest Meteorology, v. 142, no. 2–4, p. 170-185. (Also available online at *http://dx.doi.org/10.1016/j.agrformet.2006.03.026.*)

Geerts, B., 2002, On the effects of irrigation and urbanisation on the annual range of monthly-mean temperatures: Theoretical and Applied Climatology, v. 72, no. 3, p. 157-163. (Also available online at *http://dx.doi.org/10.1007/s00704-002-0683-7.*)

Gershunov, A., Cayan, D.R., and Iacobellis, S.F., 2009, The great 2006 heat wave over California and Nevada—signal of an increasing trend: Journal of Climate, v. 22, no. 23, p. 6181-6203. (Also available online at *http://dx.doi.org/10.1175/2009jcli2465.1.*)

Gillespie, A.J.R., 1999, Rationale for a national annual forest inventory program: Journal of Forestry, v. 97, no. 12, p. 16-20. (Also available online at *http://www.ingentaconnect.com/content/saf/jof/1999/00000097/00000012/art00007.*)

Global Climate Observing System (GCOS), 2010, Implementation plan for the global observing system for climate in support of the UNFCCC: Geneva, Switzerland, World Meteorological Organization, 180 p. (Also available online at *http://www.wmo.int/pages/prog/gcos/index.php?name=Publications.*)

Gordon, L.J., Steffen, W., Jönsson, B.F., Folke, C., Falkenmark, M., and Johannessen, Å., 2005, Human modification of global water vapor flows from the land surface: Proceedings of the National Academy of Sciences of the United States of America, v. 102, no. 21, p. 7612-7617. (Also available online at *http://dx.doi.org/10.1073/pnas.0500208102.*)

Guo, L.B., and Gifford, R.M., 2002, Soil carbon stocks and land use change—a meta analysis: Global Change Biology, v. 8, no. 4, p. 345-360. (Also available online at *http://dx.doi.org/10.1046/j.1354-1013.2002.00486.x.*)

Hale, R.C., Gallo, K.P., and Loveland, T.R., 2008, Influences of specific land use/land cover conversions on climatological normals of near-surface temperature: Journal of Geophysical Research-Atmospheres, v. 113, no. D14, p. D14113. (Also available online at *http://dx.doi.org/10.1029/2007jd009548.*)

Hale, R.C., Gallo, K.P., Owen, T.W., and Loveland, T.R., 2006, Land use/land cover change effects on temperature trends at U.S. Climate Normals stations: Geophysical Research Letters, v. 33, no. 11, p. L11703. (Also available online at *http://dx.doi.org/10.1029/2006gl026358.*)

Hallegatte, S., 2009, Strategies to adapt to an uncertain climate change: Global Environmental Change, v. 19, no. 2, p. 240-247. (Also available online at *http://dx.doi.org/10.1016/j.gloenvcha.2008.12.003.*)

Hand, L.M., and Shepherd, J.M., 2009, An investigation of warm-season spatial rainfall variability in Oklahoma City—possible linkages to urbanization and prevailing wind: Journal of Applied Meteorology and Climatology, v. 48, no. 2, p. 251-269. (Also available online at *http://dx.doi.org/10.1175/2008jamc2036.1.*)

Hanesiak, J.M., Raddatz, R.L., and Lobban, S., 2004, Local initiation of deep convection on the Canadian prairie provinces: Boundary-Layer Meteorology, v. 110, no. 3, p. 455-470. (Also available online at *http://dx.doi.org/10.1023/B:BOUN.0000007242.89023.e5.*)

Herrick, J.E., Lessard, V.C., Spaeth, K.E., Shaver, P.L., Dayton, R.S., Pyke, D.A., Jolley, L., and Goebel, J.J., 2010, National ecosystem assessments supported by scientific and local knowledge: Frontiers in Ecology and the Environment, v. 8, no. 8, p. 403-408. (Also available online at *http://dx.doi.org/10.1890/100017.*)

Homer, C., Huang, C., Yang, L., Wylie, B., and Coan, M., 2004, Development of a 2001 National Land Cover Database for the United States: Photogrammetric Engineering and Remote Sensing, v. 70, no. 7, p. 829-840. (Also available online at *http://asprs.org/Photogrammetric-Engineering-and-Remote-Sensing/PE-RS-Journals.html.*)

Houghton, J.T., Jenkins, G.J., and Ephrums, J.J., 1990, Climate change—the IPCC scientific assessment: Cambridge, U.K., Cambridge University Press, 410 p. (Also available online at *http://www.ipcc.ch/publications_and_data/publications_ipcc_first_assessment_1990_wg1.shtml#.T7T 56vWrRk4.*)

Howden, N.J.K., Burt, T.P., Worrall, F., Mathias, S., and Whelan, M.J., 2011, Nitrate pollution in intensively farmed regions—what are the prospects for sustaining high-quality groundwater?: Water Resources Research, v. 47, p. W00L02. (Also available online at *http://dx.doi.org/10.1029/2011wr010843.*)

Howden, S.M., Soussana, J.F., Tubiello, F.N., Chhetri, N., Dunlop, M., and Meinke, H., 2007, Adapting agriculture to climate change: Proceedings of the National Academy of Sciences of the United States of America, v. 104, no. 50, p. 19691-19696. (Also available online at *http://dx.doi.org/10.1073/pnas.0701890104.*)

Huang, C., Goward, S.N., Masek, J.G., Thomas, N., Zhu, Z., and Vogelmann, J.E., 2010, An automated approach for reconstructing recent forest disturbance history using dense Landsat time series stacks: Remote Sensing of Environment, v. 114, no. 1, p. 183-198. (Also available online at *http://dx.doi.org/10.1016/j.rse.2009.08.017.*)

Imhoff, M.L., Zhang, P., Wolfe, R.E., and Bounoua, L., 2010, Remote sensing of the urban heat island effect across biomes in the continental USA: Remote Sensing of Environment, v. 114, no. 3, p. 504-513. (Also available online at *http://dx.doi.org/10.1016/j.rse.2009.10.008.*)

Intergovernmental Panel on Climate Change (IPCC), 2007, Summary for policymakers, *in* Metz, B., ed., Climate change 2007—mitigation, contribution of Working Group III to the Fourth Assessment Report of the Intergovernmental Panel on Climate Change: Cambridge, U.K., Cambridge University Press, p. 1-23. (Also available online at *http://www.ipcc.ch/pdf/assessment-report/ar4/wg3/ar4-wg3-spm.pdf.*)

Jin, J., and Miller, N.L., 2011, Regional simulations to quantify land use change and irrigation impacts on hydroclimate in the California Central Valley: Theoretical and Applied Climatology, v. 104, no. 3-4, p. 429-442. (Also available online at *http://dx.doi.org/10.1007/s00704-010-0352-1.*)

Jobbágy, E.G., and Jackson, R.B., 2000, The vertical distribution of soil organic carbon and its relation to climate and vegetation: Ecological Applications, v. 10, no. 2, p. 423-436. (Also available online at *http://dx.doi.org/10.1890/1051-0761(2000)010[0423:tvdoso]2.0.co;2.*)

Johnson, W.C., Werner, B., Guntenspergen, G.R., Voldseth, R.A., Millett, B., Naugle, D.E., Tulbure, M., Carroll, R.W.H., Tracy, J., and Olawsky, C., 2010, Prairie wetland complexes as landscape functional units in a changing climate: BioScience, v. 60, no. 2, p. 128-140. (Also available online at *http://dx.doi.org/10.1525/bio.2010.60.2.7.*)

Junkermann, W., Hacker, J., Lyons, T., and Nair, U., 2009, Land use change suppresses precipitation: Atmospheric Chemistry and Physics, v. 9, no. 17, p. 6531-6539, available only online at *http://dx.doi.org/10.5194/acp-9-6531-2009.*

Justice, C.O., Townshend, J.R.G., Vermote, E.F., Masuoka, E., Wolfe, R.E., Saleous, N., Roy, D.P., and Morisette, J.T., 2002, An overview of MODIS land data processing and product status: Remote Sensing of Environment, v. 83, no. 1–2, p. 3-15. (Also available online at *http://dx.doi.org/10.1016/s0034-4257(02)00084-6.*)

Kalnay, E., and Cai, M., 2003, Corrigendum—impact of urbanization and land-use change on climate: Nature, v. 423, no. 102, p. 528-531. (Also available online *at http://dx.doi.org/10.1038/nature01952.*)

Kampe, T.U., Johnson, B.R., Kuester, M., and Keller, M., 2010, NEON—the first continental-scale ecological observatory with airborne remote sensing of vegetation canopy biochemistry and structure: Journal of Applied Remote Sensing, v. 4, p. 043510-043524. (Also available online at *http://dx.doi.org/10.1117/1.3361375.*)

Kaufmann, R.K., Seto, K.C., Schneider, A., Liu, Z., Zhou, L., and Wang, W., 2007, Climate response to rapid urban growth—evidence of a human-induced precipitation deficit: Journal of Climate, v. 20, no. 10, p. 2299-2306. (Also available online at *http://dx.doi.org/10.1175/jcli4109.1.*)

Kennedy, R.E., Yang, Z., and Cohen, W.B., 2010, Detecting trends in forest disturbance and recovery using yearly Landsat time series—1. LandTrendr — Temporal segmentation algorithms: Remote Sensing of Environment, v. 114, no. 12, p. 2897-2910. (Also available online at *http://dx.doi.org/10.1016/j.rse.2010.07.008.*)

Kinney, E.L., and Valiela, I., 2011, Nitrogen loading to Great South Bay—land use, sources, retention, and transport from land to bay: Journal of Coastal Research, v. 27, no. 4, p. 672-686. (Also available online at *http://dx.doi.org/10.2112/JCOASTRES-D-09-00098.1.*)

Kishtawal, C.M., Niyogi, D., Tewari, M., Pielke Sr., R.A., and Shepherd, J.M., 2010, Urbanization signature in the observed heavy rainfall climatology over India: International Journal of Climatology, v. 30, no. 13, p. 1908-1916. (Also available online at *http://dx.doi.org/10.1002/joc.2044.*)

Knorn, J., Rabe, A., Radeloff, V.C., Kuemmerle, T., Kozak, J., and Hostert, P., 2009, Land cover mapping of large areas using chain classification of neighboring Landsat satellite images: Remote Sensing of Environment, v. 113, no. 5, p. 957-964. (Also available online at *http://dx.doi.org/10.1016/j.rse.2009.01.010.*)

Kochendorfer, J.P., and Hubbart, J.A., 2010, The roles of precipitation increases and rural land-use changes in streamflow trends in the upper Mississippi River basin: Earth Interactions, v. 14, no. 20, p. 1-12. (Also available online at *http://dx.doi.org/10.1175/2010ei316.1.*)

Kueppers, L.M., Snyder, M.A., and Sloan, L.C., 2007, Irrigation cooling effect—regional climate forcing by land-use change: Geophysical Research Letters, v. 34, no. 3, p. L03703. (Also available online at *http://dx.doi.org/10.1029/2006gl028679.*)

Kueppers, L.M., Snyder, M.A., Sloan, L.C., Cayan, D., Jin, J., Kanamaru, H., Kanamitsu, M., Miller, N.L., Tyree, M., Du, H., and Weare, B., 2008, Seasonal temperature responses to land-use change in the western United States: Global and Planetary Change, v. 60, no. 3–4, p. 250-264. (Also available online at *http://dx.doi.org/10.1016/j.gloplacha.2007.03.005.*)

Kumar, S., Merwade, V., Lee, W., Zhao, L., and Song, C., 2010, Hydroclimatological impact of century-long drainage in midwestern United States—CCSM sensitivity experiments: Journal of Geophysical Research-Atmospheres, v. 115, no. D14, p. D14105. (Also available online at *http://dx.doi.org/10.1029/2009jd013228.*)

Landsberg, H.E., 1970, Man-made climatic changes: Science, v. 170, no. 3964, p. 1265-1274. (Also available online at *http://dx.doi.org/10.1126/science.170.3964.1265.*)

Lebow, Beth, Patel-Weynand, Toral, Loveland, Thomas, Guildin, Richard, and Cantral, Ralph, 2011, National climate assessment—Land use land cover national stakeholder workshop technical report: Salt Lake City, Utah, November 9 to December 1, 2011.

Le Quere, C., Raupach, M.R., Canadell, J.G., Marland, G., and et al., 2009, Trends in the sources and sinks of carbon dioxide: Nature Geoscience, v. 2, no. 12, p. 831-836. (Also available online at *http://dx.doi.org/10.1038/ngeo689.*)

Lee, E., Chase, T.N., Rajagopalan, B., Barry, R.G., Biggs, T.W., and Lawrence, P.J., 2009, Effects of irrigation and vegetation activity on early Indian summer monsoon variability: International Journal of Climatology, v. 29, no. 4, p. 573-581. (Also available online at *http://dx.doi.org/10.1002/joc.1721.*)

Lei, M., Niyogi, D., Kishtawal, C., Pielke Sr., R.A., Beltrán-Przekurat, A., Nobis, T.E., and Vaidya, S.S., 2008, Effect of explicit urban land surface representation on the simulation of the 26 July 2005 heavy rain event over Mumbai, India: Atmospheric Chemistry and Physics, v. 8, no. 20, p. 5975-5995, available only online at *http://dx.doi.org/10.5194/acp-8-5975-2008.*

Levasseur, A., Lesage, P., Margni, M., Deschênes, L., and Samson, R., 2010, Considering time in LCA—dynamic LCA and its application to global warming impact assessments: Environmental Science and Technology, v. 44, no. 8, p. 3169-3174. (Also available online at *http://dx.doi.org/10.1021/es9030003.*)

Littell, J.S., Peterson, D.L., Millar, C.I., and O'Halloran, K.A., 2012, U.S. national forests adapt to climate change through science-management partnerships: Climatic Change, v. 110, no. 1-2, p. 269-296. (Also available online at *http://dx.doi.org/10.1007/s10584-011-0066-0.*)

Liu, Y., Stanturf, J., and Goodrick, S., 2010, Trends in global wildfire potential in a changing climate: Forest Ecology and Management, v. 259, no. 4, p. 685-697. (Also available online at *http://dx.doi.org/10.1016/j.foreco.2009.09.002.*)

Lobell, D.B., Bala, G., Bonfils, C., and Duffy, P.B., 2006a, Potential bias of model projected greenhouse warming in irrigated regions: Geophysical Research Letters, v. 33, no. 13, p. L13709. (Also available online at *http://dx.doi.org/10.1029/2006gl026770.*)

Lobell, D.B., Bala, G., and Duffy, P.B., 2006b, Biogeophysical impacts of cropland management changes on climate: Geophysical Research Letters, v. 33, no. 6, p. L06708. (Also available online at *http://dx.doi.org/10.1029/2005gl025492.*)

Lobell, D.B., and Bonfils, C., 2008, The effect of irrigation on regional temperatures—a spatial and temporal analysis of trends in California, 1934–2002: Journal of Climate, v. 21, no. 10, p. 2063-2071. (Also available online at *http://dx.doi.org/10.1175/2007jcli1755.1.*)

Long, J., Nelson, T., and Wulder, M., 2010, Regionalization of landscape pattern indices using multivariate cluster analysis: Environmental Management, v. 46, no. 1, p. 134-142. (Also available online at *http://dx.doi.org/10.1007/s00267-010-9510-6.*)

Loveland, T.R., Sohl, T.L., Stehman, S.V., Gallant, A.L., Sayler, K.L., and Napton, D.E., 2002, A strategy for estimating the rates of recent United States land cover changes: Photogrammetric Engineering and Remote Sensing v. 68, no. 10, p. 1091-1099. (Also available online at *http://asprs.org/Photogrammetric-Engineering-and-Remote-Sensing/PE-RS-Journals.html.*)

Lyons, T.J., 2002, Clouds prefer native vegetation: Meteorology and Atmospheric Physics, v. 80, no. 1, p. 131-140. (Also available online at *http://dx.doi.org/10.1007/s007030200020.*)

Lyons, T.J., Nair, U.S., and Foster, I.J., 2008, Clearing enhances dust devil formation: Journal of Arid Environments, v. 72, no. 10, p. 1918-1928. (Also available online at *http://dx.doi.org/10.1016/j.jaridenv.2008.05.009.*)

Lyons, T.J., Xinmei, H., Schwerdtfeger, P., Hacker, J.M., Foster, I.J., and Smith, R.C.G., 1993, Land–atmosphere interaction in a semiarid region—the bunny fence experiment: Bulletin of the American Meteorological Society, v. 74, no. 7, p. 1327-1334. (Also available online at *http://dx.doi.org/10.1175/1520-0477(1993)074<1327:liiasr>2.0.co;2.*)

Mahmood, R., Foster, S.A., Keeling, T., Hubbard, K.G., Carlson, C., and Leeper, R., 2006, Impacts of irrigation on 20th century temperature in the northern Great Plains: Global and Planetary Change, v. 54, no. 1–2, p. 1-18. (Also available online at *http://dx.doi.org/10.1016/j.gloplacha.2005.10.004.*)

Mahmood, R., and Hubbard, K.G., 2002, Effect of time of temperature observation and estimation of daily solar radiation for the northern Great Plains, USA: Agronomy Journal, v. 94, no. 4, p. 723-733. (Also available online at *https://www.agronomy.org/publications/aj/abstracts/94/4/723.*)

Mahmood, R., and Hubbard, K.G., 2004, An analysis of simulated long-term soil moisture data for three land uses under contrasting hydroclimatic conditions in the northern Great Plains: Journal of Hydrometeorology, v. 5, no. 1, p. 160-179. (Also available online at *http://dx.doi.org/10.1175/1525-7541(2004)005<0160:aaosls>2.0.co;2.*)

Mahmood, R., Hubbard, K.G., and Carlson, C., 2004, Modification of growing-season surface temperature records in the northern great plains due to land-use transformation—verification of modelling results and implication for global climate change: International Journal of Climatology, v. 24, no. 3, p. 311-327. (Also available online at *http://dx.doi.org/10.1002/joc.992.*)

Mahmood, R., Hubbard, K.G., Leeper, R.D., and Foster, S.A., 2008, Increase in near-surface atmospheric moisture content due to land use changes—evidence from the observed dewpoint temperature data: Monthly Weather Review, v. 136, no. 4, p. 1554-1561. (Also available online at *http://dx.doi.org/10.1175/2007mwr2040.1.*)

Matyas, C., and Carleton, A., 2010, Surface radar-derived convective rainfall associations with midwest US land surface conditions in summer seasons 1999 and 2000: Theoretical and Applied Climatology, v. 99, no. 3, p. 315-330. (Also available online at *http://dx.doi.org/10.1007/s00704-009-0144-7.*)

Mawdsley, J.R., O'Malley, R., and Ojima, D.S., 2009, A review of climate-change adaptation strategies for wildlife management and biodiversity conservation: Conservation Biology, v. 23, no. 5, p. 1080-1089. (Also available online at *http://dx.doi.org/10.1111/j.1523-1739.2009.01264.x.*)

McAlpine, C.A., Syktus, J., Deo, R.C., Lawrence, P.J., McGowan, H.A., Watterson, I.G., and Phinn, S.R., 2007, Modeling the impact of historical land cover change on Australia's regional climate: Geophysical Research Letters, v. 34, no. 22, p. L22711. (Also available online at *http://dx.doi.org/10.1029/2007gl031524.*)

McAlpine, C.A., Syktus, J., Ryan, J.G., Deo, R.C., McKeon, G.M., McGowan, H.A., and Phinn, S.R., 2009, A continent under stress—interactions, feedbacks and risks associated with impact of modified land cover on Australia's climate: Global Change Biology, v. 15, no. 9, p. 2206-2223. (Also available online at *http://dx.doi.org/10.1111/j.1365-2486.2009.01939.x.*)

McDonald, R.I., Fargione, J., Kiesecker, J., Miller, W.M., and Powell, J., 2009, Energy sprawl or energy efficiency—climate policy impacts on natural habitat for the United States of America: PLoS ONE, v. 4, no. 8, p. 1-11, available only online at *http://dx.doi.org/10.1371/journal.pone.0006802.*

McGuire, A.D., Hayes, D.J., Kicklighter, D.W., Manizza, M., Zhuang, Q., Chen, M., Follows, M.J., Gurney, K.R., McClelland, J.W., Melillo, J.M., Peterson, B.J., and Prinn, R.G., 2010, An analysis of the carbon balance of the arctic basin from 1997 to 2006: Tellus, Series B: Chemical and Physical Meteorology, v. 62, no. 5, p. 455-474. (Also available online *at http://dx.doi.org/10.1111/j.1600-0889.2010.00497.x.*)

McMahon, G., Benjamin, S.P., Clarke, K., Findley, J.E., Fisher, R.N., Graf, W.L., Gundersen, L.C., Jones, J.W., Loveland, T.R., Roth, K.S., Usery, E.L., and Wood, N.J., 2005, Geography for a changing world—a science strategy for the geographic research of the U.S. Geological Survey, 2000-2015: U.S. Geological Survey Circular 1281, 54 p. (Also available online at *http://pubs.er.usgs.gov/publication/cir1281.*)

McPherson, R.A., and Stensrud, D.J., 2005, Influences of a winter wheat belt on the evolution of the boundary layer: Monthly Weather Review, v. 133, no. 8, p. 2178-2199. (Also available online at *http://dx.doi.org/10.1175/mwr2968.1.*)

McPherson, R.A., Stensrud, D.J., and Crawford, K.C., 2004, The impact of Oklahoma's winter wheat belt on the mesoscale environment: Monthly Weather Review, v. 132, no. 2, p. 405-421. (Also available online at *http://dx.doi.org/10.1175/1520-0493(2004)132<0405:tiooww>2.0.co;2.*)

Meyer, M.D., 2010, Greenhouse gas and climate change assessment: Journal of the American Planning Association, v. 76, no. 4, p. 402-412. (Also available online at *http://dx.doi.org/10.1080/01944363.2010.504808.*)

Mitra, C., Shepherd, J.M., and Jordan, T., 2011, On the relationship between the premonsoonal rainfall climatology and urban land cover dynamics in Kolkata City, India: International Journal of Climatology. (Also available online at *http://dx.doi.org/10.1002/joc.2366.*)

Mote, T.L., Lacke, M.C., and Shepherd, J.M., 2007, Radar signatures of the urban effect on precipitation distribution—a case study for Atlanta, Georgia: Geophysical Research Letters, v. 34, no. 20, p. L20710. (Also available online at *http://dx.doi.org/10.1029/2007gl031903.*)

Nagy, R.C., and Lockaby, B.G., 2011, Urbanization in the southeastern United States—socioeconomic forces and ecological responses along an urban-rural gradient: Urban Ecosystems, v. 14, no. 1, p. 71-86. (Also available online at *http://dx.doi.org/10.1007/s11252-010-0143-6.*)

Nair, U.S., Wu, Y., Kala, J., Lyons, T.J., Pielke Sr., R.A., and Hacker, J.M., 2011, The role of land use change on the development and evolution of the west coast trough, convective clouds, and precipitation in southwest Australia: Journal of Geophysical Research-Atmospheres, v. 116, no. D7, p. D07103. (Also available online at *http://dx.doi.org/10.1029/2010jd014950.*)

Nakicenovic, N., and Swart, R., eds., 2000, Special report on emissions scenarios—a special report of Working Group III of the Intergovernmental Panel on Climate Change: Cambridge, U.K., Cambridge University Press, 570 p. (Also available online at *http://www.ipcc.ch/ipccreports/sres/emission/index.php?idp=0.*)

Narisma, G.T., and Pitman, A.J., 2004, The effect of including biospheric responses to CO_2 on the impact of land-cover change over Australia: Earth Interactions, v. 8, no. 5, p. 1-28. (Also available online at *http://dx.doi.org/10.1175/1087-3562(2004)008<0001:teoibr>2.0.co;2.*)

Narisma, G.T., and Pitman, A.J., 2006, Exploring the sensitivity of the Australian climate to regional land-cover-change scenarios under increasing CO_2 concentrations and warmer temperatures: Earth Interactions, v. 10, no. 7, p. 1-27. (Also available online at *http://dx.doi.org/10.1175/ei154.1.*)

National Agricultural Statistical Service (NASS), 2009, 2007 census of agriculture, v. 1, U.S. summary and state reports: Washington, D.C., U.S. Department of Agriculture, National Agricultural Statistics Service, 639 p. (Also available online at *http://www.agcensus.usda.gov/Publications/2007/Full_Report/.*)

National Drought Mitigation Center (NDMC), 2012, VegDRI animated time series: National Drought Mitigation Center Web page at *http://vegdri.unl.edu/TimeSeries.aspx.* (Accessed May 9, 2012.)

National Oceanic and Atmospheric Administration (NOAA), 2012, National Weather Service Glossary: National Oceanic and Atmospheric Administration Web page at *http://weather.gov/glossary/.* (Accessed May 1, 2012.)

National Research Council (NRS), 2001, Grand challenges in environmental sciences: Washington, D.C., National Academy Press, 96 p. (Also available online at *http://www.nap.edu/catalog/9975.html.*)

Nelson, E., Mendoza, G., Regetz, J., Polasky, S., Tallis, H., Cameron, D.R., Chan, K.M.A., Daily, G.C., Goldstein, J., Kareiva, P.M., Lonsdorf, E., Naidoo, R., Ricketts, T.H., and Shaw, M.R., 2009, Modeling multiple ecosystem services, biodiversity conservation, commodity production, and tradeoffs at landscape scales: Frontiers in Ecology and the Environment, v. 7, no. 1, p. 4-11. (Also available online at *http://dx.doi.org/10.1890/080023.*)

Niyogi, D., Holt, T., Zhong, S., Pyle, P.C., and Basara, J., 2006, Urban and land surface effects on the 30 July 2003 mesoscale convective system event observed in the southern Great Plains: Journal of Geophysical Research-Atmospheres, v. 111, no. D19, p. D19107. (Also available online at *http://dx.doi.org/10.1029/2005jd006746.*)

Niyogi, D., Kishtawal, C., Tripathi, S., and Govindaraju, R.S., 2010, Observational evidence that agricultural intensification and land use change may be reducing the Indian summer monsoon rainfall: Water Resources Research, v. 46, no. 3, p. W03533. (Also available online at *http://dx.doi.org/10.1029/2008wr007082.*)

Niyogi, D., Pyle, P., Lei, M., Arya, S.P., Kishtawal, C.M., Shepherd, M., Chen, F., and Wolfe, B., 2011, Urban modification of thunderstorms—an observational storm climatology and model case study for the Indianapolis urban region: Journal of Applied Meteorology and Climatology, v. 50, no. 5, p. 1129-1144. (Also available online at *http://dx.doi.org/10.1175/2010JAMC1836.1.*)

Nuñez, M.N., Ciapessoni, H.H., Rolla, A., Kalnay, E., and Cai, M., 2008, Impact of land use and precipitation changes on surface temperature trends in Argentina: Journal of Geophysical Research-Atmospheres, v. 113, no. D6, p. D06111. (Also available online at *http://dx.doi.org/10.1029/2007jd008638.*)

Nusser, S.M., and Goebel, J.J., 1997, The National Resources Inventory—a long-term multi-resource monitoring programme: Environmental and Ecological Statistics, v. 4, no. 3, p. 181-204. (Also available online at *http://dx.doi.org/10.1023/a:1018574412308.*)

Oke, T.R., 1987, Boundary layer climates (2d ed.): London, U.K., John Wiley & Sons, 435 p.

Ozdogan, M., Rodell, M., Beaudoing, H.K., and Toll, D.L., 2010, Simulating the effects of irrigation over the United States in a land surface model based on satellite-derived agricultural data: Journal of Hydrometeorology, v. 11, no. 1, p. 171-184. (Also available online at *http://dx.doi.org/10.1175/2009jhm1116.1.*)

Palmer, M.A., Lettenmaier, D.P., Poff, N.L., Postel, S.L., Richter, B., and Warner, R., 2009, Climate change and river ecosystems—protection and adaptation options: Environmental Management, v. 44, no. 6, p. 1053-1068. (Also available online at *http://dx.doi.org/10.1007/s00267-009-9329-1.*)

Pan, Y., Birdsey, R., Hom, J., and McCullough, K., 2009, Separating effects of changes in atmospheric composition, climate and land-use on carbon sequestration of U.S. mid-Atlantic temperate forests: Forest Ecology and Management, v. 259, no. 2, p. 151-164. (Also available online at *http://dx.doi.org/10.1016/j.foreco.2009.09.049.*)

Parton, W.J., Morgan, J.A., Kelly, R.H., and Ojima, D.S., 2001, Modeling C responses to environmental change in grassland systems, *in* Follett, R.F., Kimble, J.M., and Lal, R., eds., The potential of U.S. grazing lands to sequester carbon and mitigate the greenhouse effect: Boca Raton, Fla., Lewis Publishers, CRC Press, p. 371-398. (Also available online at *http://www.crcnetbase.com/isbn/9781566705547.*)

Parton, W.J., Schimel, D.S., Cole, C.V., and Ojima, D.S., Analysis of factors controlling soil organic matter levels in Great Plains grasslands: Soil Science Society of America Journal, v. 51, no. 5, p. 1173-1179. (Also available online at *http://dx.doi.org/10.2136/sssaj1987.03615995005100050015x.*)

Pitman, A.J., de Noblet-Ducoudré, N., Cruz, F.T., Davin, E.L., Bonan, G.B., Brovkin, V., Claussen, M., Delire, C., Ganzeveld, L., Gayler, V., van den Hurk, B.J.J.M., Lawrence, P.J., van der Molen, M.K., Müller, C., Reick, C.H., Seneviratne, S.I., Strengers, B.J., and Voldoire, A., 2009, Uncertainties in climate responses to past land cover change—first results from the LUCID intercomparison study: Geophysical Research Letters, v. 36, no. 14, p. L14814. (Also available online at *http://dx.doi.org/10.1029/2009gl039076.*)

Polebitski, A.S., Palmer, R.N., and Waddell, P., 2011, Evaluating water demands under climate change and transitions in the urban environment: Journal of Water Resources Planning and Management, v. 137, no. 3, p. 249-257. (Also available online at *http://dx.doi.org/10.1061/(ASCE)WR.1943-5452.0000112.*)

Pongratz, J., Reick, C.H., Raddatz, T., and Claussen, M., 2010, Biogeophysical versus biogeochemical climate response to historical anthropogenic land cover change: Geophysical Research Letters, v. 37, no. 8, p. L08702. (Also available online at *http://dx.doi.org/10.1029/2010gl043010.*)

Praskievicz, S., and Chang, H., 2009, A review of hydrological modelling of basin-scale climate change and urban development impacts: Progress in Physical Geography, v. 33, no. 5, p. 650-671. (Also available online at *http://dx.doi.org/10.1177/0309133309348098.*)

Price, K., 2011, Effects of watershed topography, soils, land use, and climate on baseflow hydrology in humid regions—a review: Progress in Physical Geography, v. 35, no. 4, p. 465-492. (Also available online at *http://dx.doi.org/10.1177/0309133311402714.*)

Price, K., Jackson, C.R., Parker, A.J., Reitan, T., Dowd, J., and Cyterski, M., 2011, Effects of watershed land use and geomorphology on stream low flows during severe drought conditions in the southern Blue Ridge Mountains, Georgia and North Carolina, United States: Water Resources Research, v. 47, p. W02516. (Also available online at *http://dx.doi.org/10.1029/2010WR009340.*)

Puma, M.J., and Cook, B.I., 2010, Effects of irrigation on global climate during the 20th century: Journal of Geophysical Research-Atmospheres, v. 115, no. D16, p. D16120. (Also available online at *http://dx.doi.org/10.1029/2010jd014122.*)

Raddatz, R.L., 2003, Agriculture and tornadoes on the Canadian prairies—potential impact of increasing atmospheric CO_2 on summer severe weather: Natural Hazards, v. 29, no. 2, p. 113-122. (Also available online at *http://dx.doi.org/10.1023/a:1023626806353.*)

Raddatz, R.L., 2007, Evidence for the influence of agriculture on weather and climate through the transformation and management of vegetation—illustrated by examples from the Canadian prairies: Agricultural and Forest Meteorology, v. 142, no. 2–4, p. 186-202. (Also available online at *http://dx.doi.org/10.1016/j.agrformet.2006.08.022.*)

Raddatz, R.L., and Cummine, J.D., 2003, Inter-annual variability of moisture flux from the prairie agro-ecosystem—impact of crop phenology on the seasonal pattern of tornado days: Boundary-Layer Meteorology, v. 106, no. 2, p. 283-295. (Also available online at *http://dx.doi.org/10.1023/a:1021117925505.*)

Rashford, B.S., Walker, J.A., and Bastian, C.T., 2011, Economics of grassland conversion to cropland in the Prairie Pothole region: Conservation Biology, v. 25, no. 2, p. 276-284. (Also available online at *http://dx.doi.org/10.1111/j.1523-1739.2010.01618.x.*)

Reichstein, M., 2008, Impacts of climate change on forest soil carbon—principles, factors, models, uncertainties, *in* Freer-Smith, P.H., Broadmeadow, M.S.J., and Linch, J.M., eds., Forestry and climate change: Wallingford, U.K., CABI Books, p. 127-135.

Rose, L.S., Stallins, J.A., and Bentley, M.L., 2008, Concurrent cloud-to-ground lightning and precipitation enhancement in the Atlanta, Georgia (United States), urban region: Earth Interactions, v. 12, no. 11, p. 1-30. (Also available online at *http://dx.doi.org/10.1175/2008ei265.1.*)

Roy, D.P., Ju, J., Kline, K., Scaramuzza, P.L., Kovalskyy, V., Hansen, M., Loveland, T.R., Vermote, E., and Zhang, C., 2010, Web-enabled Landsat Data (WELD)—Landsat ETM+ composited mosaics of the conterminous United States: Remote Sensing of Environment, v. 114, no. 1, p. 35-49. (Also available online at *http://dx.doi.org/10.1016/j.rse.2009.08.011.*)

Rusticucci, M., and Barrucand, M., 2004, Observed trends and changes in temperature extremes over Argentina: Journal of Climate, v. 17, no. 20, p. 4099-4107. (Also available online at *http://dx.doi.org/10.1175/1520-0442(2004)017<4099:otacit>2.0.co;2.*)

Saeed, F., Hagemann, S., and Jacob, D., 2009, Impact of irrigation on the South Asian summer monsoon: Geophysical Research Letters, v. 36, no. 20, p. L20711. (Also available online at *http://dx.doi.org/10.1029/2009gl040625.*)

Sandstrom, M., Lauritsen, R., and Changnon, D., 2004, A central-U.S. summer extreme dew-point climatology (1949-2000): Physical Geography, v. 25, no. 3, p. 191-207. (Also available online at *http://dx.doi.org/10.2747/0272-3646.25.3.191.*)

Schlenker, W., and Roberts, M.J., 2009, Nonlinear temperature effects indicate severe damages to U.S. crop yields under climate change: Proceedings of the National Academy of Sciences, v. 106, no. 37, p. 15594-15598. (Also available online at *http://dx.doi.org/10.1073/pnas.0906865106.*)

Sen Roy, S., Mahmood, R., Niyogi, D., Lei, M., Foster, S.A., Hubbard, K.G., Douglas, E.M., and Pielke Sr., R.A., 2007, Impacts of the agricultural Green Revolution-induced land use changes on air temperatures in India: Journal of Geophysical Research-Atmospheres, v. 112, no. D21, p. D21108. (Also available online at *http://dx.doi.org/10.1029/2007JD008834.*)

Sen Roy, S., Mahmood, R., Quintanar, A., and Gonzalez, A., 2011, Impacts of irrigation on dry season precipitation in India: Theoretical and Applied Climatology, v. 104, no. 1, p. 193-207. (Also available online at *http://dx.doi.org/10.1007/s00704-010-0338-z.*)

Shem, W., and Shepherd, M., 2009, On the impact of urbanization on summertime thunderstorms in Atlanta—two numerical model case studies: Atmospheric Research, v. 92, no. 2, p. 172-189. (Also available online at *http://dx.doi.org/10.1016/j.atmosres.2008.09.013.*)

Shepherd, J.M., 2006, Evidence of urban-induced precipitation variability in arid climate regimes: Journal of Arid Environments, v. 67, no. 4, p. 607-628. (Also available online at *http://dx.doi.org/10.1016/j.jaridenv.2006.03.022.*)

Shepherd, J.M., Pierce, H., and Negri, A.J., 2002, Rainfall modification by major urban areas— observations from spaceborne rain radar on the TRMM satellite: Journal of Applied Meteorology, v. 41, no. 7, p. 689-701. (Also available online at *http://dx.doi.org/10.1175/1520-0450(2002)041<0689:rmbmua>2.0.co;2.*)

Shepherd, M., Carter, M., Manyin, M., Messen, D., and Burian, S., 2010, The impact of urbanization on current and future coastal precipitation—a case study for Houston: Environment and Planning B—Planning and Design, v. 37, no. 2, p. 284-304. (Also available online at *http://dx.doi.org/10.1068/b34102t.*)

Smith, P., Bhogal, A., Edgington, P., Black, H., Lilly, A., Barraclough, D., Worrall, F., Hillier, J., and Merrington, G., 2010, Consequences of feasible future agricultural land-use change on soil organic carbon stocks and greenhouse gas emissions in Great Britain: Soil Use and Management, v. 26, no. 4, p. 381-398. (Also available online at *http://dx.doi.org/10.1111/j.1475-2743.2010.00283.x.*)

Smith, W.B., Miles, P.D., Perry, C.H., and Pugh, S.A., 2009, Forest resources of the United States, 2007: U.S. Department of Agriculture, Forest Service, General Technical Report WO-78, 336 p., available only online at *http://www.nrs.fs.fed.us/pubs/7334.*

Sohl, Terry, and Sayler, Kristi., 2008, Using the FORE-SCE model to project land-cover change in the southeastern United States: Ecological Modelling, v. 219, no. 1–2, p. 49-65. (Also available online at *http://dx.doi.org/10.1016/j.ecolmodel.2008.08.003.*)

Solomon, B.D., 2010, Biofuels and sustainability: Annals of the New York Academy of Sciences, v. 1185, p. 119-134. (Also available online at *http://dx.doi.org/10.1111/j.1749-6632.2009.05279.x.*)

Sorooshian, S., Li, J., Hsu, K.-l., and Gao, X., 2011, How significant is the impact of irrigation on the local hydroclimate in California's Central Valley? Comparison of model results with ground and remote-sensing data: Journal of Geophysical Research-Atmospheres, v. 116, no. D6, p. D06102. (Also available online at *http://dx.doi.org/10.1029/2010jd014775.*)

Souch, C., and Grimmond, S., 2006, Applied climatology—urban climate: Progress in Physical Geography, v. 30, no. 2, p. 270-279. (Also available online at *http://dx.doi.org/10.1191/0309133306pp484pr.*)

Stallins, J.A., and Rose, L.S., 2008, Urban lightning—current research, methods, and the geographical perspective: Geography Compass, v. 2, no. 3, p. 620-639. (Also available online at *http://dx.doi.org/10.1111/j.1749-8198.2008.00110.x.*)

Stehman, S.V., and Selkowitz, D.J., 2010, A spatially stratified, multi-stage cluster sampling design for assessing accuracy of the Alaska (USA) National Land Cover Database (NLCD): International Journal of Remote Sensing, v. 31, no. 7, p. 1877-1896. (Also available online at *http://dx.doi.org/10.1080/01431160902927945.*)

Stone, B., Hess, J.J., and Frumkin, H., 2010, Urban form and extreme heat events—are sprawling cities more vulnerable to climate change than compact cities?: Environmental Health Perspectives, v. 118, no. 10, p. 1425-1428. (Also available online at *http://dx.doi.org/10.1289/ehp.0901879.*)

Tang, G., and Beckage, B., 2010, Projecting the distribution of forests in New England in response to climate change: Diversity and Distributions, v. 16, no. 1, p. 144-158. (Also available online at *http://dx.doi.org/10.1111/j.1472-4642.2009.00628.x.*)

Thompson, W.L., Miller, A.E., Mortenson, D.C., and Woodward, A., 2011, Developing effective sampling designs for monitoring natural resources in Alaskan national parks—an example using simulations and vegetation data: Biological Conservation, v. 144, no. 5, p. 1270-1277. (Also available online at *http://dx.doi.org/10.1016/j.biocon.2010.09.032.*)

Tomer, M.D., and Schilling, K.E., 2009, A simple approach to distinguish land-use and climate-change effects on watershed hydrology: Journal of Hydrology, v. 376, no. 1-2, p. 24-33. (Also available online at *http://dx.doi.org/10.1016/j.jhydrol.2009.07.029.*)

Trenberth, K.E., 2011, Challenges in GEWEX, International Global Energy and Water Cycle Experiment project office: GEWEX News, v. 21, no. 4, p. 2-3. (Also available online at *http://www.gewex.org/gewex_nwsltr.html.*)

Trusilova, K., Jung, M., Churkina, G., Karstens, U., Heimann, M., and Claussen, M., 2008, Urbanization impacts on the climate in Europe—numerical experiments by the PSU–NCAR Mesoscale Model (MM5): Journal of Applied Meteorology and Climatology, v. 47, no. 5, p. 1442-1455. (Also available online at *http://dx.doi.org/10.1175/2007jamc1624.1.*)

Turner, B.L., Kasperson, R.E., Matson, P.A., McCarthy, J.J., Corell, R.W., Christensen, L., Eckley, N., Kasperson, J.X., Luers, A., Martello, M.L., Polsky, C., Pulsipher, A., and Schiller, A., 2003, A framework for vulnerability analysis in sustainability science: Proceedings of the National Academy of Sciences, v. 100, no. 14, p. 8074-8079. (Also available online at *http://dx.doi.org/10.1073/pnas.1231335100.*)

Turner, B.L., Lambin, E.F., and Reenberg, A., 2007, The emergence of land change science for global environmental change and sustainability: Proceedings of the National Academy of Sciences, v. 104, no. 52, p. 20666-20671. (Also available online at *http://dx.doi.org/10.1073/pnas.0704119104.*)

U.S. Environmental Protection Agency (US EPA), 2009, Land-use scenarios—national-scale housing-density scenarios consistent with climate change storylines (final report): EPA/600/R-08/076F, 137 p., available only online at *http://cfpub.epa.gov/ncea/cfm/recordisplay.cfm?deid=203458.*

U.S. Environmental Protection Agency (US EPA), 2012, Green power defined: U.S. Environmental Protection Agency Web page at *http://www.epa.gov/greenpower/gpmarket/index.htm.* (Accessed May 1, 2012.)

U.S. Geological Survey (USGS), 2012, Land Cover Trends Project: U.S. Geological Survey database available online at *http://landcovertrends.usgs.gov/download/dlMap.html.* (Accessed May 2, 2012.)

U.S. Global Change Research Program, 2000, US National Assessment of the Potential Consequences of Climate Variability and Change Sector: Agriculture: U.S. Global Change Research Program, accessed May 16, 2012 available at *http://www.usgcrp.gov/usgcrp/Library/nationalassessment/15AG.pdf.*

United Nations Framework Convention on Climate Change (UNFCCC), 2012a, Land use, land-use change and forestry: Draft Decision /CMP.6, 9 p. (Also available online at *http://unfccc.int/files/meetings/cop_16/application/pdf/cop16_lulucf.pdf.*)

United Nations Framework Convention on Climate Change (UNFCCC), 2012b, Outcome of the work of the ad hoc working group on long-term cooperative action under the convention: Draft Decision /CP.16, 29 p. (Also available online at *http://unfccc.int/files/meetings/cop_16/application/pdf/cop16_lca.pdf.*)

van den Heever, S.C., and Cotton, W.R., 2007, Urban aerosol impacts on downwind convective storms: Journal of Applied Meteorology and Climatology, v. 46, no. 6, p. 828-850. (Also available online at *http://dx.doi.org/10.1175/jam2492.1.*)

Verburg, P.H., Neumann, K., and Nol, L., 2011, Challenges in using land use and land cover data for global change studies: Global Change Biology, v. 17, no. 2, p. 974-989. (Also available online at *http://dx.doi.org/10.1111/j.1365-2486.2010.02307.x.*)

Viger, R.J., Hay, L.E., Markstrom, S.L., Jones, J.W., and Buell, G.R., 2011, Hydrologic effects of urbanization and climate change on the Flint River basin, Georgia: Earth Interactions, v. 15, no. 20, p. 1-25. (Also available online at *http://dx.doi.org/10.1175/2010ei369.1.*)

Vogelmann, J.E., Kost, J.R., Tolk, B., Howard, S., Short, K., Chen, X., Huang, C., Pabst, K., and Rollins, M.G., 2011, Monitoring landscape change for LANDFIRE using multi-temporal satellite imagery and ancillary data: IEEE Journal of Selected Topics in Applied Earth Observations and Remote Sensing, v. 4, no. 2, p. 252-264. (Also available online at *http://dx.doi.org/10.1109/JSTARS.2010.2044478.*)

Voldseth, Richard, Johnson, W. Carter, Guntenspergen, Glenn, Gilmanov, Tagir, and Millett, Bruce, 2009, Adaptation of farming practices could buffer effects of climate change on northern prairie wetlands: Wetlands, v. 29, no. 2, p. 635-647. (Also available online at *http://dx.doi.org/10.1672/07-241.1.*)

Waisanen, P.J., and Bliss, N.B., 2002, Changes in population and agricultural land in conterminous United States counties, 1790 to 1997: Global Biogeochemical Cycles, v. 16, no. 4, p. 1137. (Also available online at *http://dx.doi.org/10.1029/2001gb001843.*)

West, J.M., Julius, S.H., Kareiva, P., Enquist, C., Lawler, J.J., Petersen, B., Johnson, A.E., and Shaw, M.R., 2009, U.S. natural resources and climate change—concepts and approaches for management adaptation: Environmental Management, v. 44, no. 6, p. 1001-1021. (Also available online at *http://dx.doi.org/10.1007/s00267-009-9345-1.*)

Westerling, A.L., Turner, M.G., Smithwick, E.A.H., Romme, W.H., and Ryan, M.G., 2011, Continued warming could transform Greater Yellowstone fire regimes by mid-21st century: Proceedings of the National Academy of Sciences, v. 108, no. 32, p. 13165-13170. (Also available online at *http://dx.doi.org/10.1073/pnas.1110199108.*)

Wickham, J.D., Stehman, S.V., Fry, J.A., Smith, J.H., and Homer, C.G., 2010, Thematic accuracy of the NLCD 2001 land cover for the conterminous United States: Remote Sensing of Environment, v. 114, no. 6, p. 1286-1296. (Also available online at *http://dx.doi.org/10.1016/j.rse.2010.01.018.*)

Wimberly, M.C., Hildreth, M.B., Boyte, S.P., Lindquist, E., and Kightlinger, L., 2008, Ecological niche of the 2003 West Nile virus epidemic in the northern Great Plains of the United States: PLoS ONE, v. 3, no. 12, p. 1-7, available only online at *http://dx.doi.org/10.1371/journal.pone.0003744.*

Yadav, S.S., Redden, R.J., Hatfield, J.L., Lotze-Campen, H., and Hall, A.E., eds., 2011, Crop adaptation to climate change: Cambridge, U.K., John Wiley & Sons, 632 p. (Also available online at *http://onlinelibrary.wiley.com/book/10.1002/9780470960929.*)

York, A.M., Shrestha, M., Boone, C.G., Zhang, S., Harrington, J.A., Jr., Prebyl, T.J., Swann, A., Agar, M., Antolin, M.F., Nolen, B., Wright, J.B., and Skaggs, R., 2011, Land fragmentation under rapid urbanization—a cross-site analysis of southwestern cities: Urban Ecosystems, v. 14, no. 3, p. 429-455. (Also available online at *http://dx.doi.org/10.1007/s11252-011-0157-8.*)

Yow, D.M., 2007, Urban heat islands—observations, impacts, and adaptation: Geography Compass, v. 1, no. 6, p. 1227-1251. (Also available online at *http://dx.doi.org/10.1111/j.1749-8198.2007.00063.x.*)

Zhao, M., Pitman, A.J., and Chase, T.N., 2001, Climatic effects of land cover change at different carbon dioxide levels: Climate Research, v. 17, no. 1, p. 1-18. (Also available online at *http://dx.doi.org/10.3354/cr017001.*)

Zheng, D., Heath, L.S., Ducey, M.J., and Smith, J.E., 2011, Carbon changes in conterminous US forests associated with growth and major disturbances—1992-2001 (vol 6, 0140212, 2011): Environmental Research Letters, v. 6, no. 1, p. 1, available only online at *http://dx.doi.org/10.1088/1748-9326/6/1/019502.*

Zhou, Yan, and Shepherd, J. Marshall, 2010, Atlanta's urban heat island under extreme heat conditions and potential mitigation strategies: Natural Hazards, v. 52, no. 3, p. 639-668. (Also available online at *http://dx.doi.org/10.1007/s11069-009-9406-z.*)

Zhu, Zhiliang, ed., Bouchard, Michelle, Butman, David, Hawbaker, Todd, Li, Zhengpeng, Liu, Jinxun, Liu, Shuguang, McDonald, Cory, Reker, Ryan, Sayler, Kristi, Sleeter, Benjamin, Sohl, Terry, Stackpoole, Sarah, Wein, Anne, and Zhu, Zhiliang, 2011, Baseline and projected future carbon storage and greenhouse-gas fluxes in the Great Plains region of the United States: U.S. Geological Survey Professional Paper 1787, 28 p. (Also available online at *http://pubs.er.usgs.gov/publication/pp1787.*)

Zhu, Zhiliang, ed., Bergamaschi, Brian, Bernknopf, Richard, Clow, David, Dye, Dennis, Faulkner, Stephen, Forney, William, Gleason, Robert, Hawbaker,Todd, Liu, Jinxun, Liu, Shuguang, Prisley, Stephen, Reed, Bradley, Reeves, Matthew, Rollins, Matthew, Sleeter, Benjamin, Sohl, Terry, Stackpoole, Sarah, Stehman, Stephen, Striegl, Robert, Wein, Anne, and Zhu, Zhiliang, 2012, A method for assessing carbon stocks, carbon sequestration, and greenhouse-gas fluxes in ecosystems of the United States under present conditions and future scenarios: U.S. Geological Survey Scientific Investigations Report 2010-5233, 190 p. (Also available online at *http://pubs.er.usgs.gov/publication/sir20105233*.)

Appendix A – Definitions of Major Terms

Land change: The patterns, processes, and consequences of changes in land use, land condition, and land cover at multiple spatial and temporal scales that is associated with the interaction between human activities and natural systems. (U.S. Geological Survey Geography Science Sythesis Team, written commun., 2010).

Land use: The human utilization of environmental resources for economic, societal, or environmental development.

Land cover: The observed biophysical characteristics of the Earth's surface.

Land condition: The observable state of the Earth's surface used for monitoring and assessing resilience and sustainability as a function of both natural and human processes.

Weather: The state of the atmosphere in a given geographic region and specific point in time with respect to its effects on human-environmental systems and variables such as temperature and moisture. (Adapted from National Oceanic and Atmospheric Administration, 2012 and National Aeronautics and Space Administration, 2012).

Climate: The long-term, average weather pattern for a given region.

Risk: The probability of a consequence from a known or forecasted change in climate or land cover, use, or condition.

Vulnerability: The degree to which a coupled human-environmental system is susceptible to reduced function, goods, and services because of adverse effects of climate change, including climate variability and extremes. Vulnerability is a function of the exposure, sensitivity, and adaptive capacity of the system given the character, magnitude, and rate of climate change. (Krista Karstensen and Nathan Wood, U.S. Geological Survey, written commun., 2012).

Appendix B. Land Cover Class Descriptions

Water – Areas persistently covered with water, such as streams, canals, lakes, reservoirs, bays, or oceans.

Developed – Areas of intensive use with much of the land covered with structures or anthropogenic impervious surfaces (for example, high-density residential, commercial, industrial, roads) or less intensive uses where the land-cover matrix includes both vegetation and structures (for example, low-density residential, recreational facilities, cemeteries, parking lots, utility corridors), including any land functionally related to urban or built-up environments (for example, parks, golf courses).

Mining – Areas with extractive mining activities that have a substantial surface expression. This includes (to the extent that these features can be detected) mining buildings, quarry pits, overburden, leach, evaporative, tailings, or other related components.

Barren – Land comprised of soils, sand, or rocks where less than 10 percent of the area is vegetated. Barren lands are usually naturally occurring.

Forest – Tree-covered land where the tree cover density is greater than 10 percent. Note that cleared forest land (clear-cuts) is mapped according to current cover (for example, mechanically disturbed or grassland/shrubland).

Grassland/Shrubland – Land predominately covered with grasses, forbs, or shrubs. The vegetated cover must comprise at least 10 percent of the area.

Agriculture – Land in either a vegetated or an unvegetated state used for the production of food and fiber. This includes cultivated and uncultivated croplands, hay lands, pasture, orchards, vineyards, and confined livestock operations. Note that forest plantations are considered forests regardless of the use of the wood products.

Wetland – Land where water saturation is the determining factor in soil characteristics, vegetation types, and animal communities. Wetlands usually contain water and vegetated cover.

Ice and Snow – Land where the accumulation of snow and ice does not completely melt in the summer period (for example, alpine glaciers and snowfields).

Nonmechanically disturbed – Land in an altered and often unvegetated state that because of disturbances by nonmechanical means, is in transition from one cover type to another. Nonmechanical disturbances are caused by fire, wind, floods, animals, and other similar phenomena.

Mechanically disturbed – Land in an altered and often unvegetated state that, because of disturbances by mechanical means, is in transition from one cover type to another. Mechanical disturbances include forest clear-cutting, earthmoving, scraping, chaining, reservoir drawdown, and other similar human-induced changes.

Appendix C. 1973–2000 National Climate Assessment Southeast Region Change Statistics

Percentage of land cover classes in the Southeast

	Water	Developed	Mechanically Disturbed	Mining	Barren	Forest	Grassland	Agriculture	Wetland	Nonmechanically disturbed	Snow/Ice
1973	4.58	5.99	1.35	0.33	0.05	48.26	0.85	27.17	11.40	0.01	0.00
1980	4.73	6.45	1.66	0.41	0.05	47.27	0.96	27.27	11.18	0.01	0.00
1986	4.86	6.90	1.97	0.41	0.04	46.69	1.14	26.96	10.99	0.03	0.00
1992	4.93	7.45	2.56	0.36	0.04	46.37	1.23	26.08	10.99	0.01	0.00
2000	5.04	8.27	3.11	0.28	0.04	45.75	1.16	25.55	10.71	0.09	0.00
1973-2000	0.45	2.28	1.76	-0.05	-0.01	-2.51	0.31	-1.62	-0.69	0.07	0.00

Gross and net change in the Southeast

	Gross (percent change)	Net (percent change)
Water	0.64	0.45
Developed	2.30	2.28
Mechanically Disturbed	3.90	1.76
Mining	0.47	-0.05
Barren	0.03	-0.01
Forest	7.91	-2.51
Grassland/Schrubland	0.91	0.31
Agriculture	3.67	-1.62
Wetland	1.31	-0.69
Nonmechanically disturbe	0.10	0.07
Snow/Ice	0.00	0.00

Leading land cover conversions in the Southeast from 1973-2000

Rank	From	To	Change (square kilometers)	Standard error (square kilometers)	Change (percent of region)	Standard error (percent of region)
1	Forest	Mechanically disturbed	34,220	3,323	2.50	0.24
2	Agriculture	Forest	20,317	3,290	1.49	0.24
3	Forest	Developed	16,422	2,359	1.20	0.17
4	Mechanically disturbed	Forest	12,868	1,761	0.94	0.13
5	Forest	Agriculture	11,661	1,149	0.85	0.08
6	Agriculture	Developed	10,759	1,479	0.79	0.11
7	Wetland	Water	4,271	2,002	0.31	0.15
8	Forest	Grassland/Shrubland	3,768	636	0.28	0.05
9	Wetland	Mechanically disturbed	3,175	624	0.23	0.05
10	Wetland	Developed	2,586	1,146	0.19	0.08

Appendix D. 1973–2000 National Climate Assessment Northwest Region Change Statistics

Percentage of land cover classes in the Northwest

	Water	Developed	Mechanically Disturbed	Mining	Barren	Forest	Grassland	Agriculture	Wetland	Nonmechanically disturbed	Snow/Ice
1973	1.34	1.16	1.35	0.04	1.53	42.72	36.93	13.66	0.85	0.17	0.25
1980	1.32	1.27	1.12	0.05	1.53	42.27	37.44	13.90	0.84	0.01	0.25
1986	1.38	1.37	1.30	0.05	1.52	41.88	37.21	13.90	0.85	0.28	0.25
1992	1.27	1.51	1.76	0.06	1.52	41.29	37.57	13.34	0.88	0.56	0.25
2000	1.32	1.67	1.42	0.07	1.53	40.34	37.27	13.31	0.88	1.95	0.25
1973-2000	-0.02	0.51	0.07	0.03	0.00	-2.39	0.35	-0.35	0.03	1.78	0.00

Gross and net change in the Northwest

	Gross (percent change)	Net (percent change)
Water	0.12	-0.02
Developed	0.51	0.51
Mechanically Disturbed	2.71	0.07
Mining	0.05	0.03
Barren	0.05	0.00
Forest	6.04	-2.39
Grassland/Schrubland	4.58	0.35
Agriculture	1.94	-0.35
Wetland	0.12	0.03
Nonmechanically disturbed	2.10	1.78
Snow/Ice	0.00	0.00

Leading land cover conversions in the Northwest from 1973-2000

Rank	From	To	Change (square kilometers)	Standard error (square kilometers)	Change (percent of region)	Standard error (percent of region)
1	Forest	Grassland/Shrubland	8,869	952	1.38	0.15
2	Forest	Mechanically disturbed	8,681	852	1.35	0.13
3	Mechanically disturbed	Forest	7,450	1,009	1.16	0.16
4	Forest	Nonmechanically disturbed	7,380	2,863	1.15	0.45
5	Agriculture	Grassland/Shrubland	5,460	1,238	0.85	0.19
6	Grassland/Shrubland	Nonmechanically disturbed	5,022	2,558	0.78	0.40
7	Grassland/Shrubland	Agriculture	4,622	1,247	0.72	0.19
8	Grassland/Shrubland	Forest	3,311	457	0.52	0.07
9	Forest	Developed	1,595	222	0.25	0.03
10	Agriculture	Developed	1,280	331	0.20	0.05

Appendix E. 1973–2000 National Climate Assessment Great Plains Region Change Statistics

Percentage of land cover classes in the Great Plains

	Water	Developed	Mechanically Disturbed	Mining	Barren	Forest	Grassland/Shrubland	Agriculture	Wetland	Nonmechanically disturbed	Snow/Ice
1973	1.95	0.99	0.24	0.06	0.74	12.61	48.26	33.55	1.53	0.06	0.01
1980	1.94	1.11	0.24	0.09	0.74	12.47	47.96	33.91	1.54	0.00	0.01
1986	1.99	1.19	0.42	0.10	0.73	12.21	47.86	33.98	1.51	0.00	0.01
1992	1.95	1.28	0.35	0.11	0.75	12.00	49.52	32.24	1.54	0.23	0.01
2000	2.19	1.42	0.34	0.13	0.75	11.90	49.00	31.95	1.39	0.12	0.01
1973-2000	0.23	0.43	0.11	0.07	0.00	0.71	1.55	1.60	0.13	0.06	0.00

Gross and net change in the Great Plains

	Gross (percent change)	Net (percent change)
Water	0.46	0.23
Developed	0.43	0.43
Mechanically Disturbed	0.52	0.11
Mining	0.09	0.07
Barren	0.09	0.00
Forest	1.50	-0.71
Grassland/Schrubland	5.29	1.55
Agriculture	4.82	-1.60
Wetland	0.40	-0.13
Nonmechanically disturbed	0.18	0.06
Snow/Ice	0.00	0.00

Leading land cover conversions in the Great Plains from 1973-2000

Rank	From	To	Change (square kilometers)	Standard error (square kilometers)	Change (percent of region)	Standard error (percent of region)
1	Agriculture	Grassland/Shrubland	63,299	5,196	2.75	0.23
2	Grassland/Shrubland	Agriculture	32,723	2,295	1.42	0.10
3	Forest	Grassland/Shrubland	10,686	1,975	0.46	0.09
4	Forest	Mechanically disturbed	6,320	770	0.27	0.03
5	Agriculture	Developed	4,870	1,044	0.21	0.05
6	Wetland	Water	4,809	796	0.21	0.03
7	Forest	Agriculture	3,503	357	0.15	0.02
8	Grassland/Shrubland	Forest	3,369	424	0.15	0.02
9	Mechanically disturbed	Forest	3,042	547	0.13	0.02
10	Grassland/Shrubland	Developed	3,030	745	0.13	0.03

Appendix F. 1973–2000 National Climate Assessment Northeast Region Change Statistics

Percentage of land cover classes in the Northeast

	Water	Developed	Mechanically Disturbed	Mining	Barren	Forest	Grassland/Shrubland	Agriculture	Wetland	Nonmechanically disturbed	Snow/Ice
1973	4.93	8.45	0.41	0.40	0.08	63.83	0.37	18.07	3.46	0.00	0.00
1980	4.94	8.71	0.62	0.41	0.08	63.27	0.59	17.93	3.44	0.00	0.00
1986	4.95	8.97	0.86	0.42	0.08	62.63	0.87	17.77	3.43	0.02	0.00
1992	4.96	9.33	1.04	0.42	0.08	62.14	1.11	17.50	3.42	0.01	0.00
2000	4.96	9.81	1.08	0.54	0.08	61.81	1.10	17.22	3.41	0.00	0.00
1973-2000	0.03	1.36	0.66	0.14	0.00	-2.02	0.73	-0.85	-0.05	0.00	0.00

Gross and net change in the Northeast

	Gross (percent change)	Net (percent change)
Water	0.06	0.03
Developed	1.37	1.36
Mechanically Disturbed	1.42	0.66
Mining	0.53	0.14
Barren	0.01	0.00
Forest	4.05	-2.02
Grassland/Schrubland	1.14	0.73
Agriculture	1.48	-0.85
Wetland	0.08	-0.05
Nonmechanically disturbed	0.00	0.00
Snow/Ice	0.00	0.00

Leading land cover conversions in the Northeast from 1973-2000

Rank	From	To	Change (square kilometers)	Standard error (square kilometers)	Change (percent of region)	Standard error (percent of region)
1	Forest	Mechanically disturbed	5,351	835	1.00	0.16
2	Forest	Developed	3,911	391	0.73	0.07
3	Forest	Grassland/Shrubland	3,838	979	0.71	0.18
4	Agriculture	Developed	3,170	410	0.59	0.08
5	Agriculture	Forest	2,165	232	0.40	0.04
6	Mechanically disturbed	Forest	1,826	335	0.34	0.06
7	Forest	Mining	1,589	458	0.30	0.09
8	Forest	Agriculture	1,531	180	0.28	0.03
9	Grassland/Shrubland	Forest	961	254	0.18	0.05
10	Agriculture	Grassland/Shrubland	674	135	0.13	0.03

Appendix G. 1973–2000 National Climate Assessment Midwest Region Change Statistics

Percentage of land cover classes in the Midwest

	Water	Developed	Mechanically Disturbed	Mining	Barren	Forest	Grassland/Shrubland	Agriculture	Wetland	Nonmechanically disturbed	Snow/Ice
1973	3.45	3.56	0.26	0.15	0.02	28.33	1.81	56.21	6.22	0.00	0.00
1980	3.45	3.82	0.25	0.15	0.02	28.04	1.93	56.08	6.20	0.05	0.00
1986	3.50	4.02	0.40	0.15	0.02	27.78	2.03	55.94	6.16	0.00	0.00
1992	3.50	4.31	0.44	0.17	0.02	27.56	2.32	55.50	6.17	0.00	0.00
2000	3.54	4.89	0.57	0.17	0.02	27.40	2.41	54.82	6.17	0.01	0.00
1973-2000	0.08	1.34	0.32	0.02	0.00	-0.93	0.59	-1.38	-0.05	0.01	0.00

Gross and net change in the Midwest		
	Gross (percent change)	Net (percent change)
Water	0.25	0.08
Developed	1.35	1.34
Mechanically Disturbed	0.81	0.32
Mining	0.13	0.02
Barren	0.01	0.00
Forest	2.26	-0.93
Grassland/Schrubland	1.35	0.59
Agriculture	2.74	-1.38
Wetland	0.34	-0.05
Nonmechanically distur	0.01	0.01
Snow/Ice	0.00	0.00

Leading land cover conversions in the Midwest from 1973-2000

Rank	From	To	Change (square kilometers)	Standard error (square kilometers)	Change (percent of region)	Standard error (percent of region)
1	Agriculture	Developed	12,571	1,700	1.04	0.14
2	Agriculture	Grassland/Shrubland	7,817	1,261	0.65	0.10
3	Forest	Mechanically disturbed	6,311	883	0.52	0.07
4	Forest	Agriculture	5,889	477	0.49	0.04
5	Forest	Grassland/Shrubland	3,206	555	0.27	0.05
6	Forest	Developed	2,815	364	0.23	0.03
7	Agriculture	Forest	2,676	260	0.22	0.02
8	Mechanically disturbed	Forest	2,380	517	0.20	0.04
9	Grassland/Shrubland	Forest	2,332	364	0.19	0.03
10	Grassland/Shrubland	Agriculture	1,630	414	0.14	0.03

Appendix H. 1973–2000 National Climate Assessment Southwest Region Change Statistics

Percentage of land cover classes in the Southwest

	Water	Developed	Mechanically Disturbed	Mining	Barren	Forest	Grassland/Shrubland	Agriculture	Wetland	Nonmechanically disturbed	Snow/Ice
1973	0.47	1.08	0.08	0.20	2.08	22.19	66.65	6.55	0.64	0.05	0.01
1980	0.54	1.21	0.06	0.23	2.08	22.10	66.31	6.72	0.61	0.13	0.01
1986	0.53	1.31	0.12	0.25	2.08	22.08	66.18	6.78	0.61	0.06	0.01
1992	0.50	1.46	0.20	0.27	2.08	21.98	66.54	6.26	0.61	0.08	0.01
2000	0.50	1.59	0.15	0.30	2.08	21.70	66.36	6.18	0.61	0.51	0.01
1973-2000	0.03	0.51	0.07	0.10	0.00	-0.49	-0.28	-0.37	-0.02	0.46	0.00

Gross and net change in the Southwest

	Gross (percent change)	Net (percent change)
Water	0.13	0.03
Developed	0.52	0.51
Mechanically Disturbed	0.22	0.07
Mining	0.12	0.10
Barren	0.02	0.00
Forest	0.75	-0.49
Grassland/Schrubland	2.51	-0.28
Agriculture	1.73	-0.37
Wetland	0.12	-0.02
Nonmechanically disturb	0.55	0.46
Snow/Ice	0.00	0.00

Leading land cover conversions in the Southwest from 1973-2000

Rank	From	To	Change (square kilometers)	Standard error (square kilometers)	Change (percent of region)	Standard error (percent of region)
1	Agriculture	Grassland/Shrubland	14,766	3,389	0.82	0.19
2	Grassland/Shrubland	Agriculture	11,895	1,887	0.66	0.10
3	Forest	nmechanically disturb	5,393	1,843	0.30	0.10
4	Grassland/Shrubland	Developed	5,289	1,200	0.29	0.07
5	Grassland/Shrubland	nmechanically disturb	3,702	1,323	0.21	0.07
6	Forest	Grassland/Shrubland	3,523	521	0.20	0.03
7	Agriculture	Developed	3,504	1,066	0.19	0.06
8	Forest	Mechanically disturbed	1,727	401	0.10	0.02
9	Grassland/Shrubland	Mining	1,665	389	0.09	0.02
10	Grassland/Shrubland	Forest	1,087	256	0.06	0.01

Appendix I. Summaries from Land-Use Land-Cover and Climate Case Studies

Climate Influences on LCLU and Water Resources Case Study Summaries

Baseflow, Drought, and Flood

"The Roles of Precipitation Increases and Rural Land-Use Changes in Streamflow Trends in the Upper Mississippi River Basin" (Kochendorfer and Hubbart, 2010)

Precipitation and streamflow trends were examined for 48 United States Hydro-Climatic Data Network (HCDN) streams in the upper Mississippi water resource region from 1939 through 2008. Using the concept of precipitation of elasticity of flow, the observed magnitudes of statistically significant increases in mean and low flows were up to a factor of three greater than those expected from observed increases in precipitation alone. Peak flows increased less than expected. The authors suggest that the differences between the expected and observed changes in streamflow can be explained by regional, rural land-use changes. Results emphasized that precipitation trends alone cannot explain the direction and magnitude of flow trends citing specifically how soil and water conservation efforts have impacted rural land change. Another major conclusion applicable to land change was that all seven streams in the Driftless Area, an area with a history of widespread soil conservation drivers of land change, had decreasing peak flows despite an increase in precipitation. Because of the fact that the study streams were relatively free of significant changes in developed areas and flow regulation, the authors suggest that rural land-use change has played a significant role in streamflow alterations. Moreover, they suggest that the assumption of flow trends in HCDN streams being exclusively because of climate change is likely incorrect for the upper Mississippi River Basin and any other agriculturally dominated region in the United States.

"Effects of watershed topography, soils, land use, and climate on baseflow hydrology in humid regions: A review"(Price, 2011)

Studies on climatic and land change influences on baseflow are split. Many studies associate higher watershed forest cover with lower baseflows, attributed to high evapotranspiration (ET) rates of forests, while other studies suggest that baseflow increased with higher watershed forest cover because of higher infiltration and recharge. Moreover, the effects of specific land-cover effects (developed and agriculture) are also varied because of additions of imported water and variable background conditions. The author reviewed a variety of studies in humid, temperate, and tropical regions and suggested that there is a research need to address land and climate change effects and human impacts on baseflow.

Urbanization – There were many inconsistencies in the studies examined by the author, likely stemming from the variety of factors controlling baseflow discharge and system response to urbanization.

Agriculture – As with urbanization, baseflow response to agricultural land use may be positive or negative depending on the management practice. Despite inconsistency in the literature, two main inferences were made by the author: (1) watersheds that have been under agricultural land use for extended periods show baseflow increases in response to improved cropping and tillage practices, and (2) comparison of baseflows under agricultural land use compared to other land uses is precluded by the variety of management practices and variables uses of irrigation.

"Effects of watershed land use and geomorphology on stream low flows during severe drought conditions in the southern Blue Ridge Mountains, Georgia and North Carolina, United State." (Price and others, 2011)

The objective of this study was to assess the influence of land use and watershed geomorphic characteristics on low-flow variability in the southern Blue Ridge Mountains of North Carolina and Georgia. Two low-flow seasons, coinciding with a severe drought period in the Southeastern United States, were monitored in 35 small-scales to mesocscale watersheds 93 to 146 km^2. A comprehensive suite of watershed characteristics, including factors of topography, channel network morphometry, soil, land use, and precipitation were used in multiple regression analysis of low-flow variability within the 35 watersheds. The results of this study indicate that low flows in the southern Blue Ridge Mountains are affected most strongly by factors of geomorphology, particularly drainage density, topographic variability, amount of watershed colluvium, and percentage of the stream network that is first order. Additionally, low flow in groups of lower and higher forest-cover watersheds were compared. There is a significant body of literature that has demonstrated lower baseflow within higher watershed forest cover, this study also demonstrated a consistent positive relation between low flow and higher forest cover. This study highlights the importance of infiltration and recharge under undisturbed land cover in sustaining low flows, and bears noteworthy implications for environmental flows and water resource sustainability. These results also suggest that as development continues, additional land-use change will carry a negative impact on reduced low flow, thus issues of environmental flows will be reduced as forest is converted to nonforest developed lands.

"Hydrologic Effects of Urbanization and Climate Change on the Flint River Basin, Georgia" (Viger and others 2011)

This study used the Precipitation-Runoff Modeling System (PRMS) to examine the potential effects of long-term urbanization on the freshwater resources of the Flint River System. Climate inputs included precipitation and temperature output from five, downscaled General Circulation Models (GCMs) (one current and four future) and land cover. The central tendency of streamflow simulated in the climate-change scenarios showed a slight decrease in overall streamflow relative to simulations under current conditions, mostly caused by decreases in the surface-runoff and groundwater components. The addition of forecasted urbanization of land surfaces from the Forecasting Scenarios of Future Land-Cover Model (FORE–SCE) to the hydrologic simulation mitigated the decreases in streamflow by increasing surface runoff. The authors did note that there was a large degree of uncertainty about forecasted precipitation in interpreting the results of the study as the Flint River streamflow is extremely sensitive to precipitation adding that it may be more sensitive to precipitation than to increases in impervious surfaces.

"A simple approach to distinguish land-use and climate-change effects on watershed hydrology" (Tomer and Schilling, 2009)

This paper notes that it is difficult to separate the influences of climate (long-term averages) from weather cycles (drought and excess precipitation) and land-use change on watershed hydrology (annual hydrologic budgets). For a 25-year, small-watershed experiment in Iowa, the authors were able to separate some of these interactions. If applied regionally, the conclusions suggest that climate change has increased discharge from Midwest watersheds, especially since the 1970s. By inference, climate change has increased susceptibility of nutrients to water transport, exacerbating Gulf of Mexico hypoxia.

Water Supply

"Can forest management be used to sustain water-based ecosystem services in the face of climate change?"(Ford and others, 2011)

Forested watersheds provide ecosystem services related to water supply and can have their structure, function, and streamflow altered by land use and land cover. The goal of this study was to quantify interactions among forest management, climate, and streamflow. It was hypothesized that climate impacts may be mitigated or exacerbated by forest management practices that alter land cover, depending upon how land cover changes impact hydrologic functions. A retrospective analysis and synthesis of long-term climate and streamflow data from six watersheds, all with different management histories, were studied to determine whether streamflow responded to variation in annual temperature and extreme precipitation compared to unmanaged watersheds. It was found that the streamflow response to climate was affected by nearly all land uses examined. Watershed response differed significantly in extreme wet and dry years. Deciduous stands that were converted to pine altered the streamflow response to extreme wet years. The increase of soil water storage may reduce flood risks in wet years but could exacerbate drought during a dry year. Forest management could potentially mitigate extreme annual precipitation associated with climate change; however, management effects need to be analyzed spatially to determine potential risks and vulnerabilities.

"A review of hydrological modeling of basin-scale climate change and urban development impacts" (Praskievicz and Chang, 2009)

This study reviewed the components of hydrologic modeling of climate change and urbanization at the basin scale. The most significant effect of increasing urbanization is the increase in impervious surfaces that contributes to increases in surface runoff, flashiness of storm hydrographs, degradation of water quality, and decreased groundwater recharge and evapotranspiration. For geographic and land use and land cover relevance, the authors discuss how the projected impacts of climate change on basin hydrology depend on the geography of the study area (that is, arid regions are likely to experience a decrease in annual runoff), how basin size may influence hydrologic sensitivity to urban development (that is, smaller basins experiencing greater impacts than larger ones), and the location of the development (that is, increased impervious surfaces in headwater regions tend to have more impact than development downstream). There are studies that suggest that land-use change will affect water resources more significantly than climate change (Cuo and others, 2009), and there are studies that suggest future climate changes are more significant than land-use change in determining hydrological impacts (Choi, 2008).

For vulnerability and water management related studies, the authors suggest that climate change is likely to exacerbate conflicts over competing land-uses for water resources (that is, urban and agricultural). This problem maybe the most severe in regions where projected increased temperatures would diminish snowpack and shift the timing of peak runoff to earlier in the spring, which could lead to lower flows in the summer when water demand is generally high (that is, mountains in the Western United States).

"Evaluating water demands under climate change and transitions in the urban environment" (Polebitski and others, 2011)

This study discusses how population growth, land use, pricing policy, and climate change affect residential water demands in the Puget Sound Region. The authors used a water demand model coupled with an urban simulation model (UrbanSim) to generate water demands over a 30-year time span (2000–30) using four different scenarios (including a baseline).

Scenario 1: Demonstrated the importance of incorporating urban growth and development in determining water demands especially in single-family homes. The authors suggest that changes in development patterns, such as conversion of larger single-family lots into smaller townhomes can have a significant influence on water demand.

Scenarios 2 and 3: The scenarios influenced the summer demand differently. While Seattle has one of the Nation's lowest outdoor watering rates, the city still consumes about 1.4 times more water during the summer than in the winter. In the scenarios, the water pricing policy decreases the demand whereas increasing temperatures increase the demand. The authors note that the increases in summer water consumption because of climate change may be ameliorated by pricing policies and conservation policies, and offer as support that Seattle has been experiencing a decreasing water demand over the past two decades because of conservation programs, responses to price increases, increasing fixture efficiency, and two curtailment periods.

An interesting perspective offered in the conclusion of this study was that housing density may prove to be essential in understanding effects of climate change on water demands – a factor that is currently not incorporated into many forecast models.

Water Quality

"Stream discharge and riparian land use influence in-stream concentrations and loads of phosphorus from central plains watershed." (Banner and others, 2009)

This study investigated linkages between land-cover use and discharge of total phosphorus (TP) concentration loads in 25 Kansas streams. A long-term monitoring dataset maintained by the Kansas Department of Health and Environment was used to extract data from collection points that were colocated with USGS gaging stations. Median TP concentration was strongly correlated ($R^2 = 76$ percent) with the presence of cropland within the riparian zones. This study found that on average 88 percent of the TP loading occurred during high flows, which occurred 10 percent of the time. The authors suggest that in order to reduce TP during baseflow conditions, management activities may be best focused on reducing cropland uses near perennial streams.

"A century of changing land-use and water-quality relationships in the continental US" (Broussard and Turner, 2009)

The relationships between various land-use practices and riverine nitrate-nitrogen (NN) concentrations from the early 1990s through the end of the 20th century were studied in order to determine if there was a statistical significance in the quantifiable association between agricultural land-use and riverine NN concentrations. It was found that 63 rivers monitored over the past century for NN concentrations were three to four times higher at the end of the century. The most noticeable change of increased NN concentrations was in the Mississippi River Basin and the Midwest region. Agricultural cropland planted with corn was the most important agricultural driver of riverine NN concentrations, according to the factors considered in this study. This study predicts that the continued expansion of modern corn agriculture will likely increase the NN concentration in rivers and streams, particularly in watersheds where corn cropland occupies more than 25 percent of the total area. The authors suggest that change in land-use practices, such as increasing cropland diversity or increasing the area of perennial crops could help improve water quality by reducing NN loading.

"Nitrate in groundwater of the United States, 1991–2003" (Burow and others, 2010)

An increase in fertilizer usage has been necessary for increasing crop production for food to support an increasing world population. In the United States, the amount of Nitrogen(N)entering the environment from the application of N fertilizer has increased 20-fold since 1945. With groundwater supplying 33 percent of water used for public drinking, contamination of this water by increased nitrates can affect human health. The objective of this paper was to identify the most influential factors controlling groundwater nitrate concentrations across a wide range of environmental setting in the United States. This study used an extensive data set from the USGS National Water-Quality Assessment (NAWQA). During 1991 through 2003, 5,101 wells were sampled in 51 study areas throughout the United States. The NAWQA method for assessing groundwater focuses on the water-quality conditions of shallow groundwater beneath agricultural and urban land-use settings and in major aquifers in each study area. Nitrate concentrations were highest in shallow groundwater beneath agricultural land use in areas with well-drained soils and oxic geochemical conditions. Nitrate concentrations were lowest in deep groundwater where ground water is reduced, or where groundwater is older and, hence, concentrations reflect historically low N application rates.

"Nitrogen loading to Great South Bay: Land use, sources, retention, and transport from land to bay" (Kinney and Valiela, 2011).

A nitrogen-loading model was used to (1) estimate annual nitrogen deliverables to each subwatershed of Great South Bay (GSB), (2) partition the contribution of nitrogen through disposal of wastewater, use of fertilizers, and atmospheric deposition on land, (3) estimate nitrogen transport through major land-use categories (natural vegetation, wetlands, agriculture, tuft, impervious surfaces), (4) calculate nitrogen retention within each subwatershed, and (5) model land-derived nitrogen loads from each subwatershed to GSB. Wastewater-derived nitrogen was the dominant source to watershed surfaces followed by atmospheric deposition and fertilizer use. There was a high within watershed nitrogen retention (77 percent) linked to areas of natural vegetation. The authors concluded that as land cover shifts away from natural vegetation, eutrophication of GSB will increase.

Wetlands

"Prairie wetland complexes as landscape: Functional units in a changing climate" (Johnson and others, 2010)

The approach examined the consequences of climate change on wetland complexes in the Prairie Pothole Region (PPR) using WETLANDSCAPE (WLS), which is a climate-driven, process-based, deterministic simulation model. Generally, the simulations showed that all three permanence types of wetlands lost significant hydroperiod under warming scenarios unless accompanied by a minimum increase in precipitation of 5–7 percent per degree of warming. The model suggests that the vulnerability of the members of prairie wetland complexes to the climate warming defined by the study generally increase in the following order: temporary wetlands, semipermanent wetlands, seasonal wetlands. Principal conclusions as written by the authors are: (1) prairie wetlands in general are highly sensitive to climate warming; (2) wetlands in the drier, western PPR are most vulnerable to climate warming; (3) members of the wetland complex will respond differently to climate change, and longer-hydroperiod wetlands are perhaps the most sensitive; (4) shortened wetland hydroperiods will severely affect vertebrates because of their longer life-cycle requirements; (5) in a greenhouse climate more of the PPR will be too dry or without functional wetlands and nesting habitat to support historic levels of waterfowl breeding; and (6) adaptation of farming practices in wetland watersheds may buffer the effects of climate change on wetlands.

"Hydroclimatological impact of century-long drainage in midwestern United States: CCSM sensitivity experiments" (Kumar and others, 2010)

This study aimed to fill the research gap by analyzing the effect of wetland drainage on Midwest hydroclimatology testing two hypotheses: (1) large-scale artificial drainage has significantly changed the energy and water fluxes in the Midwest, thus affecting precipitation and temperature in the region, and (2) the impact of past LCLU change on Midwest hydroclimatology is comparable to or even greater than the impact of greenhouse gas emission-based climate change in the region. The study conducted four sensitivity experiments to investigate the effects of climate and LULC changes in the Midwest using a community climate system model (CCSM3). It is important to note that "climate change" in this study refers to greenhouse gas emission-based climate change.

Large-scale drainage activities in the Midwest (Illinois, Indiana, Iowa, Michigan, Minnesota, Missouri, Ohio, and Wisconsin) have resulted in the loss of more than 85 percent of its original wetlands. Moreover, agricultural drainage of Midwest wetlands has changed the carbon balance of the region from carbon sink to carbon source and has affected the roles wetlands play in climatic feedbacks (that is albedo and Bowen ratio). The results of the modeling suggested that wetland drainage has resulted in significant changes in the energy budget (sensible and latent heat flux) and radiation budget (long-wave radiation), particularly from May to October. As a result, the authors state that the climate has become warmer and convective precipitation has decreased during summer months. However, climate change did not result in any significant changes in other important variables (that is, precipitation, latent heat flux, long-wave radiation). Results from this study highlight the importance of LULC change compared to climate change in the Midwest over the past one and half centuries.

"Adaptation of farming practices could buffer effects of climate change on northern prairie wetlands" (Voldseth and others, 2009)

The authors used WETSIM 3.2 to examine the effects of climate and watershed cover on wetland water levels in eastern South Dakota to answer the questions: Can adaptation of agricultural land-use practices ameliorate effects of climate change on prairie wetlands? and How much climate change can be offset or absorbed through land-use management? The considered land-use practices were unmanaged grassland, managed grassland, and cultivated crops. Climate scenarios were developed by adjusting historical climate combinations of $2^{\circ}C$ and $4^{\circ}C$ air temperature and +/- 10 percent precipitation. Water levels in wetlands surrounded by managed grasslands were significantly greater than those surrounded by unmanaged grassland. Management reduced the proportion of years the wetland went dry and the frequency of dry periods. Cultivated crops and managed grasslands returned water levels that were equal or greater than unmanaged grassland under historical climate for the $2^{\circ}C$ temperature increase and the $2^{\circ}C$ and 10 percent precipitation increase. Managed grasslands returned water levels that were equal or greater than unmanaged grassland under historical climate for the $4^{\circ}C$ temperature increase and the $2^{\circ}C$ and 10 percent precipitation increase. The authors state that few empirical data exist to verify the results of such land-use simulations.

Climate Influences on Land Use and Land Cover and Transportation and Energy Supply and Use Case Study Summaries

Transportation

"Roadless areas and clean water" (DellaSala and others, 2011)

In the United States, inventoried roadless areas (IRAs) are lands without roads exceeding 2,000 ha (5,000 acres) and that have been inventoried by the USFS. This paper assessed the importance of preserving IRAs as developing them could severely impact water quality and supply. Colorado has been seeking Federal permission to develop IRAs for logging, skiing, mining, and producing oil and gas. However, IRAs in Colorado occur in the headwaters of all major drainages, which cover about a one-third of the upper watersheds in the State. If road building was allowed in these intact watersheds it could reduce the ability to provide clean water to downstream communities. It was projected that the demand for water in Colorado will triple by 2050. Development of IRAs in Colorado would provide short-term opportunities but long-term impacts on water supply and quality. The authors suggest the use of cost-benefit analyses to better inform project management on decisions that may reduce the water quality.

"Assessing transportation infrastructure impacts on rangelands: Test of a standard rangeland assessment protocol" (Duniway and others, 2010)

The increase of on- and off-road vehicles, in recent decades, is because of a variety of factors including increased availability of all-terrain vehicles, infrastructure development (oil, gas, renewable energy, and ex-urban), and recreational activities. With continued development of renewable energy sources, including wind and solar, additional service roads are anticipated to be developed in rangelands. The goal of this study was to test the application of the Interpreting Indicators of Rangeland Health (IIRH) for detecting impacts on rangelands because of roads, trails, and pipelines. Results indicate that IIRH can be used for detecting areas impacted by off-highway vehicle use and energy use. These methods could potentially be integrated with remote sensing technologies in order to assess landscape scale impacts of transportation networks.

Energy Supply and Use

"The land use-climate change-energy nexus" (Dale and others, 2011)

Land use, climate change, and energy influence each other through a complex set of processes that need to be documented, modeled, and calibrated. This study explored the use of landscape ecology on patterns and processes and how they could be used to provide insight into the interaction between climate change, energy choices, and land-use activities. Major research issues within the land use-climate change-energy nexus were discussed in order to address integrated patterns and processes. It was found that climate influences the potential output, relative efficiencies, and sustainability of alternative energy sources. Landscape ecology could be used to enhance the understanding of interactions among land use, energy, and climate change. Understanding how climate change alters landscape conditions is important because management decisions should be based on scientific study that consider both the short-term and long-term implications of climate change.

"Energy sprawl or energy efficiency: Climate policy impacts on natural habitat for the United States of America" (McDonald and others, 2009)

Climate change has been acknowledged as a potential threat to biodiversity and human well-being, which has led the United States to consider a cap-and-trade system to regulate emissions. Scenarios of a cap-and-trade system's effects on United States' energy production and impacts on habitat because of energy sprawl were evaluated. In all scenarios, temperate deciduous forests and temperate grasslands were identified as the most impacted habitats by future energy development. It was found that regardless of a cap-and-trade bill, 206,000 km^2 of land will be affected by energy production by 2030. The results demonstrated that a potential side effect to reducing emissions would increase habitat impacts. The possibility of widespread energy sprawl increases the need for energy conservation, appropriate siting, sustainable production practices, and compensatory mitigation offsets.

"Assessing transportation infrastructure impacts on rangelands: Test of a standard rangeland assessment protocol" (Duniway and others, 2010)

The increase of on- and off-road vehicles, in recent decades, is because of a variety of factors including increased availability of all-terrain vehicles, infrastructure development (oil, gas, renewable energy, and ex-urban), and recreational activities. With continued development of renewable energy sources, including wind and solar, additional service roads are expected to be developed in rangelands. The goal of this study was to test the application of the Interpreting Indicators of Rangeland Health (IIRH) for detecting impacts on rangelands because of roads, trails, and pipelines. Results indicate that IIRH can be used for detecting areas impacted by off-highway vehicle use and energy use. These methods could potentially be integrated with remote sensing technologies in order to assess landscape scale impacts of transportation networks.

"Biofuels and sustainability" (Solomon, 2010)

There has been considerable controversy in North America, Europe, and Southeast Asia on the sustainably of the biofuel industry. This study focused on ethanol manufactured from cornstarch in the United States and from sugarcane in Brazil, and biodiesel produced from rapeseed oil in Germany and France. Recent literature in ecological economics and sustainability science were reviewed and several common criteria for sustainable biofuels were identified: scale (resource assessment, land availability, and land use practices), efficiency (economic and energy), equity (geographic distribution of resources and the "food versus fuel" debate), socio-economic issues, and environmental effects and emissions. A number of analyses found that corn-based ethanol was unsustainable and had significant environmental costs. Sugarcane-based ethanol and biodiesel were difficult to determine their sustainability, because of significant environmental challenges. The only biofuel found to be produced and consumed on a sustainable basis was cellulosic ethanol. Cellulosic ethanol produces a large resource and land base for feedstock, higher energy return on investment, potentially greater economic efficiency, equitable resource distribution, little or no conflict with food resources, and lower greenhouse gas emissions. It is important to consider sustainable biofuels especially with growing concerns about food security, distribution, greenhouse gas emissions, soil erosion, water pollution, and water supply in arid regions.

Climate Influences on Land Use and Land Cover and Agriculture Case Study Summaries

Land Use Conversion to Cropland

"Changes in historical Iowa land cover as context for assessing the environmental benefits of current and future conservation efforts on agricultural lands" (Gallant and others, 2011)

Two prominent conservation programs, from the 2008 Farm Bill, that authorized the USDA to provide financial incentives for agricultural producers in order to reduce environmental impacts included the Wetlands Reserve Program (WRP) and the Conservation Reserve Program (CRP). Current and future success of WRP and CRP practices depends on incentives for landowners to enroll or stay enrolled in conservation programs. With increased prices of corn and soybeans there has been a substantial decrease in CRP throughout the Midwest. This study analyzed land-cover changes since the mid-1800s in Iowa to better understand the full range of benefits from WRP and CRP efforts in Iowa. The contemporary Iowa landscape is extensively managed for agriculture and has significantly changed from the grassland-, woodland-, and wetland-covered landscape it once was in the mid-1800s. These landscape-scale changes were linked to conversion of grasslands and forest to agriculture and draining of wetlands. The resulting land-cover analysis of Iowa provides a description of how humans have altered the landscape since the mid-1800s. The spatial and temporal changes in the amount and distribution of grasslands, woodlands, and wetlands can provide the USDA with a context to evaluate benefits of conservation programs.

"Economics of Grassland Conversion to Cropland in the Prairie Pothole Region" (Rashford and others, 2011)

Much of the remaining grassland in the Prairie Pothole Region (PPR) is privately owned, and its conversion to cultivated cropland is largely driven by economics. This study investigated land-use change in the PPR and explored the economic and landscape characteristics related to conservation. The study modeled 183 counties in Iowa, Minnesota, North Dakota, South Dakota, and Montana in order to estimate the probability of grassland conversion to cropland. Results of the model demonstrated that conversion probabilities are spatially heterogeneous as a function of soil quality and are driven by changes in economic returns of alternate land uses.

Cropland Sustainability

"Fertilizer usage has increased in order to increase crop production" (Banner and others, 2009; Broussard and Turner, 2009; Burow and others, 2010).

Increased Growing Season Temperatures

"Nonlinear temperature effects indicate severe damages to U.S. crop yield under climate change" (Schlenker and Roberts, 2009).

The study pairs a panel of county-level yields for corn, soybeans, and cotton with a new fine-scale weather dataset that incorporates the whole distribution of temperatures within each day and across all days in the growing season. The data spanned most United States counties from 1950 to 2005. Results show that yields increase with temperature up to 29°C for corn, 30°C for soybeans, and 32°C for cotton, but temperatures above these thresholds are harmful. The nonlinear relations established in this study revealed that (1) the relation between yield and temperature observed in cooler northern States is similar to that observed in southern States, and (2) the relation between yield and temperature observed between 1950 and 1977 was the same as observed between 1978 and 2005. An important caveat of the study was the inability to account for CO_2 concentrations.

Climate Influences on Land Use and Land Cover and Ecosystems Case Study Summaries

Greenhouse Gas Inventory and Biogenic Emissions

"Future land use and land cover influences on regional biogenic emissions and air quality in the United States" (Chen and others, 2009)

This study applied a regional modeling system driven by global models to study the land use and land cover effects on future biogenic emissions and their impacts on ozone and biogenic secondary organic aerosols (BSOA). Under the IPCC SRES A2 business-as-usual storyline, the modeling system was applied to a matrix of four model scenarios including: Case 1-present day 1990 through 1999, and three future cases covering 2045 through 2054; Case 2 -present day LULC; Case 3 -projected-future LULC; and Case 4 - future LULC with designed regions of tree planting for carbon sequestration. Results of the study showed changing future meteorology under Case 2 increased average isoprene and monoterpene emission rates by 26 percent and 20 percent becuse of higher temperature and solar insolation. Under Case 3 predicted isoprene and monoterpene emissions decreased by 52 percent and 31 percent respectively, primarily because of the projected expansion of cropland. The reduction of isoprene and monoterprene was less under Case 4 at 31 percent and 14 percent when future LULC changes were accompanied by regions of tree plantings. Overall, the results demonstrated that on a regional basis, changes in LULC can offset temperature driven increases in biogenic emissions and, thus, LULC projection is an important factor to consider in the study of future regional air quality.

"Carbon changes in conterminous US forests associated with growth and major disturbances: 1992–2001" (Zheng and others, 2011)

Forest ecosystems are an important part of the global carbon cycle, therefore, the United Nations Framework Convention and Climate Change (UNFCCC) agreed that qualifying nations report their greenhouse gas (GHG) emissions and sinks annually. The goal of this study was to estimate the effects of major forest disturbances and net growth on carbon sequestration in the conterminous United States in context of the terminology and needs for reporting to the UNFCCC for national GHG inventories. Forest area and carbon changes were estimated using remote-sensing based land-cover change map, forest fire data from the Monitoring Trends in Burn Severity Program, and forest growth and harvest data from the USFW Forest Inventory and Analysis Program. Natural and human-associated disturbances reduced the forest ecosystem's carbon sink by 36 percent from 1992 to 2001, compared to that without disturbances. Among the three identified disturbances (land-cover change, harvesting, and forest wildfires) forest-related land-cover change contributed 33 percent of the total effect in reducing the forest carbon potential sink, while harvests and fires accounted for 63 percent and 4 percent of the total effect, respectively. The southern region of the United States was a small net carbon source, whereas, the greater Pacific Northwest region was a strong net sink. Results from this study fit reasonably well at an aggregated level with other related estimates of the current forest GHG inventory in the United States, suggesting that further research using this approach is acceptable.

Vegetation Displacement

"Projecting the distribution of forests in New England in response to climate change" (Tang and Beckage, 2010)

Three major forest types in New England were modeled to project potential distributions in response to expected climate change. The three dominant forest types in New England include: boreal conifer, northern deciduous hardwood, and mixed oak-hickory forests. Historically, the distribution of these forest types have corresponded with a natural climate gradient. Boreal conifer forests are found at higher elevations in northern regions of New England, northern deciduous hardwoods are found in the central uplands, and mixed oak-hickory are found at lower elevations in the more southerly regions. Using the process-based BIOME4 vegetation model and nine future climate change scenarios, the future distribution of the three primary forest types were simulated to determine the potential magnitude of spatial displacement of the forest types by mid and late 21st century. The model projected that climate warming in the 21st century is likely to cause the extensive loss of boreal conifer forests, reduce the extent of northern hardwood deciduous forests, and result in large increases of mixed oak-hickory forest.

"Assessing ecosystem threats from global and regional change: Hierarchical modeling of risk to sagebrush ecosystems from climate change, land use and invasive species in Nevada, USA" (Bradley, 2010)

A lack of spatially explicit models of risk to ecosystems makes it difficult for science to inform conservation planning and management. Bradley presents a series of modeling approaches to (1) estimate the likelihood of change to climatic habitat of sagebrush, (2) develop a statewide landscape-scale risk assessment associated with land use and the invasion of cheatgrass, and (3) model risk of pinyon-juniper woodland expansion into sagebrush ecosystems. The study found that the risk of sagebrush loss associated with land use and cheatgrass invasion was highest in portions of Nevada already dominated by cheatgrass and near lands used for agriculture. Regionally, there was considerable spatial heterogeneity in future climatic suitability for sagebrush. Wyoming, eastern Idaho, central Oregon, and northern Nevada had the lowest predicted risk of climate change. In Nevada, only a small portion of the State was assessed to contain low risk sagebrush, with 97 percent of the sagebrush at risk from one or more of the factors modeled.

"Potential of a national monitoring program for forests to assess change in high-latitude ecosystems." (Barret and Gray, 2011)

This paper presents the potential benefits of extending the Forest Inventory and Analysis (FIA) approach to the boreal forest region of Alaska. The authors recognize that climate change is expected to impact many components of boreal ecosystems, but for most indicators, the direction and magnitude of change are difficult to predict because of complex interactions among system components. The authors suggest that the status and trend information provided by FIA monitoring could be helpful to conservation decisions by providing information on the abundance and rarity of vascular plants, invasive species, biomass and carbon content of vegetation, shifting vegetation species distribution, disturbance frequency, type, impact, and wildlife habitat characteristics.

Increased Wildland Fires

"Trends in global wildfire potential in a changing climate" (Liu and others, 2010)

Global wildfire potential and projected future climate change trends because of greenhouse effects were investigated. In this study fire potential was measured by the Keetch-Byram Drought Index, which was calculated using observed maximum temperature and precipitation and future climate conditions at the end of this century (2070–2100) measured by general circulation models. Future increases in fire potential were found in five regions: the United States, South America, Eurasia, southern Africa, and Australia. The increased fire potential is primarily caused by warming in mid-latitude regions of the United States, Eurasia, and Australia and by the combination of warming and drying in other regions. The overall results suggest global increases in wildfire potential, which would require increased resources and management efforts for disaster prevention and recovery.

"Continued warming could transform GreaterYellowstone fire regimes by mid-21st century" (Westerling and others, 2011)

In response to climate change, wildfire regimes, fire frequency, extent, severity, and seasonality are expected to increase. However, potential climate-driven changes in regional fire regimes are not well understood. This study considers how the occurrence, size, and spatial location of large fires greater than 200 hectares in the Greater Yellowstone ecosystem (GYE) might be altered by climate change. A suite of statistical models that related monthly climate data (1972 to 1999) to the occurrence and size of fires was developed, cross-validated, and then used with global climate models to predict fire occurrence and area burned in the GYE through 2009. It was found that continued warming could transform GYE fire regimes by the mid-21st century. The magnitude of predicted increases in fire occurrences and area burned suggested that there is a likelihood that the forested area in the GYE will be converted to nonforest vegetation and thus alter the fire return interval. The authors suggest that the climate-fire system is a tipping element that may qualitatively change the ecosystem processes in GYE and other subalpine or boreal forests.

Climate Influences on Human Health Case Study Summaries

West Nile Virus

"Weather and land cover influences on mosquito populations in Sioux Falls, South Dakota" (Chuang and others, 2011)

Spatial and temporal patterns of two mosquito populations (*Culex tarsalis* Coquillett and *Aedes vexans* Meigen) were examined with their relations to land cover types and climatic variability in Sioux Falls, S.Dak. South Dakota is dominated by two species of mosquitos, *Cx. tarsalis and Ae. vexans*, with the first a known vector of West Nile virus (WNV) and the second a suspected bridge vector of WNV. It was found that *Cx. tarsalis* was positively correlated with grass/hay, and *Ae. vexans* was positively correlated with wetlands. Seasonal weather patterns had an effect on the abundance of both mosquito species. Precipitation 2–3 weeks before mosquito sampling and warm temperature during the collection week had a positive influence on *Ae. vexans*. Warm temperatures 2 weeks prior along with increased precipitation 3–4 weeks before mosquito sampling had a positive influence on *Cx. tarsalis*. This study helped to improve the understanding of mosquito ecology in particular to weather conditions and landscape characteristics that make the Northern Great Plains an environmental niche. Understanding temporal and spatial distributions of two important mosquito species and their relations with land cover and weather may help to enhance the efficiency of vector control efforts and disease prevention.

"Ecological niche of the 2003 West Nile virus epidemic in the Northern Great Plains of the United States" (Wimberly and others, 2008)

Incidences of West Nile virus (WNV) have remained higher in the northern Great Plains than in other regions of the United States. Yet, the reasons for the sustained high risk of WNV in this region have not been determined. To assess the environmental drivers of WNV in the Great Plains, county-level spatial patterns of human cases during the 2003 epidemic were analyzed across seven States. This study found that WNV incidence increased with mean May-July temperatures and also increased with the percentage of irrigated cropland and population living in rural areas. There is evidence that the environmental conditions across the northern Great Plains create a favorable ecological niche for *Culex tarslis* Coquillett, a known vector of the WNV.

Extreme Heat Events

"The great 2006 heat wave over California and Nevada: Signal of an increasing trend" (Gershunov and others, 2009)

This case study took a comprehensive look at the July 2006 heat wave over California and Nevada in context to the region's climate over the last six decades. The great California and Nevada heat waves can be classified into daytime or nighttime events depending on whether atmospheric conditions are dry or humid. Nighttime and daytime heat waves can be distinguished by the anomaly of moist atmosphere during nighttime events. The entire region has experienced a positive trend in heat wave activity, which has been expressed strongly in nighttime temperature extremes. Recently, the last two nighttime heat waves were also strongly expressed in daytime temperatures. These changes in regional heat wave activities are related to the presence of an increasing Pacific moisture source to southwestern California, west of Baja, California. Temperature records from the last 59 years indicate that the couplings of daytime and nighttime heat waves have increased. The authors suggest that the recent increase in California and Nevada heat waves are consistent with regional signs of global warming.

"Urban form and extreme heat events: Are sprawling cities more vulnerable to climate change than compact cities?" (Stone and others, 2010)

Extreme heat events (EHEs) are becoming more frequent in large cities in the United States and are responsible for a greater annual number of climate-related fatalities. This heat effect is a phenomenon through which cities exhibit higher temperatures than the surrounding countryside, in particular, low-density, sprawling patterns of urban development. This study examined the association between urban form at the level of the metropolitan region and the frequency of EHEs over a five-decade period. A sprawl index for 53 metropolitan regions in the United Sates was used to measure the association between urban form in 2000 and the mean annual rate of change in EHEs between 1956 and 2005. It was found that the frequency of EHEs increased significantly on an annual basis in large metropolitan regions of the United States. The rate of increase in EHEs was found to be higher in more sprawling than compact metropolitan regions. The authors suggest that their findings have clear implications for public health officials and urban planners to incorporate land-use patterns into models that project climate change impacts over time in urban areas.

Climate Influences on Land Use and Land Cover and Society Case Study Summaries

"Urbanization in the Southeastern United States: Socioeconomic forces and ecological responses along and urban-rural gradient" (Nagy and Lockaby, 2011)

Results from the west Georgia (WestGA) project evaluated the causes and consequences of urbanization associated with a midsize city (less than 200,000 population). The hypothesis tested was that negative feedback would occur as a result of environmental impacts that could alter the rate of development or its spatial distribution. Collectively, the WestGA project found that urbanization had greatly altered aquatic and terrestrial ecosystems through vegetation clearing, impervious surfaces, changes in hydrology, and degradation of water quality. With increasing land values, market concentrations, road accessibility, and forest to urban conversion, it was foundthat residents of developing and urban areas were unaware of the surrounding degraded ecosystems. Consequently, the hypothesis was rejected suggesting that environmental conditions had little influences on rates and patterns of development. The authors state that quantified ecological response to development along the urban-rural gradient could help predict changes that will occur with increased urbanization in the WestGA Piedmont regions.

"Urban modification of thunderstorms: An observational storm climatology and model case study for the Indianapolis urban region" (Niyogi and others, 2011)

To understand the impact of urban area on thunderstorms, 10 years of summertime thunderstorms were analyzed to capture varying storm events that were initiated around the Indianapolis urban region and peripheral rural counties. It was hypothesized that urban regions would alter the intensity and composition /structure of approaching thunderstorms because of land-surface heterogeneity. It was found that 60 percent of storms changed structure over the Indianapolis area as compared to 25 percent over the rural regions. Radar-base climatology indicated that storms split close to the upwind urban region and merge again downwind. A larger portion of small storms and larger storms was found downwind from the urban region, and midsize storms dominated the upwind region. A case study of a typical storm on June 13, 2005, was observed, and results indicated that urban regions have strong climatological influences on regional thunderstorms.

"Land fragmentation under rapid urbanization: A cross-site analysis of southwestern cities" (York and others, 2011)

To better understand land fragmentation in the Southwestern United States, national land cover data were analyzed from 1992 to 2001 in five southwestern cities associated with Long Term Ecological Research (LTER) sites. Two general fragmentation trends were observed: (1) expansion of the urbanized area leading to fragmentation in the exurban and peri-urban regions and (2) decreased fragmentation associated with infill in the previously developed urban areas. Three fragmentation patterns that reflected western growth and urbanization were identified as riparian, polycentric, and monocentric. Five drivers of land fragmentation patterns were identified including: water provisioning, urban population dynamics, transportation, topography, and institutional factors (these factors are components within the socio-ecological system or external drivers). Results from this study identified water as the key variable in understanding land change in the Southwestern United States because all five study sites had dammed major rivers for storage or prevention of flooding.